The Unified Process Elaboration Phase

Best Practices in Implementing the UP

Scott W. Ambler and Larry L. Constantine,
Compiling Editors

Masters collection from

SOFTWARE Development

CRC Press
Taylor & Francis Group
Boca Raton London New York

CRC Press is an imprint of the
Taylor & Francis Group, an **informa** business

CRC Press
Taylor & Francis Group
6000 Broken Sound Parkway NW, Suite 300
Boca Raton, FL 33487-2742

First issued in hardback 2017

© 2000 by Taylor & Francis Group, LLC
CRC Press is an imprint of Taylor & Francis Group, an Informa business

No claim to original U.S. Government works

ISBN 13: 978-1-138-41225-5 (hbk)
ISBN 13: 978-1-929629-05-3 (pbk)

Visit the Taylor & Francis Web site at
http://www.taylorandfrancis.com

and the CRC Press Web site at
http://www.crcpress.com

Cover Art Design: Robert Ward and John Freeman

To my grandparents:
Sheldon, Frances, Vivian, and William.

Thank you.

Table of Contents

Chapter 4   Best Practices for the Requirements Workflow **95**

Chapter 5   Best Practices for the Infrastructure Management Workflow. **131**

Chapter 6   Best Practices for the Analysis and Design Workflow. **185**

Foreword

The skeleton in the closet of the Information Technology (IT) industry is that large-scale, mission-critical projects suffer from an 80 to 90 percent failure rate. I know this firsthand from having been a corporate developer myself for more than 15 years before becoming an editor over three years ago. When I first came on board *Software Development* magazine as Technical Editor, Scott Ambler was just starting to make a name for himself through his "Thinking Objectively" column. Writing in everyday language that the average developer could understand, covering topics pertinent to real-world developers, Scott was staking an expertise claim in leading-edge areas such as process patterns, using the UML effectively, and mapping objects to relational databases. Now Scott has turned his considerable interest and enthusiasm into attempting to improve the abysmal failure rate of most software projects

Software process is all about standardizing the efforts of your development team and promoting best practices that have been long recognized in the industry, such as code inspections, configuration management, change control, and architectural modeling. With his process patterns background, Scott is in an ideal position to offer a consistent and focused assessment of the object-oriented software processes currently vying for mindshare, such as the Rational Unified Process (RUP), the OPEN Process from the OPEN consortium (www.open.org), and the Object-Oriented Software Process (OOSP). Although OOSP (described in Scott's books *Process Patterns* and *More Process Patterns* from Cambridge University Press) offers better inherent support for process patterns, it is a mark of Scott's real-world industry acumen that, for better or worse, he recognizes that the Unified Process is the market leader and will likely remain so despite its challenges. As Scott says ("Enhancing the Unified Process,"

Software Development, October 1999), "The Unified Process isn't perfect — none of [the software processes] are — but it's the one that we need to make work as an industry if we want organizations to be successful at software." This series augments the barebones Unified Process with material drawn from both the OPEN Process and the OOSP. A fleshed-out Unified Process strikes me as one of the best ways to deal with skeletons in the closet, and Scott is a darn good anatomy teacher.

Roger Smith
Technical Editor
Software Development magazine

Preface

A vast wealth of knowledge has been published in *Software Development* (www.sdmaga-zine.com), and its original incarnation of *Computer Language* before it, about how to be successful at developing software. The list of people who have written for, and continue to write for, the magazine is impressive: Steve McConnell, Ed Yourdon, Larry Constantine, Steve McCarthy, Clemens Szyperski, Peter Coad, and Karl Wiegers to name a few. In short, the leading minds in the information industry have shared their wisdom with us over the years in the pages of this venerable magazine.

Lately, there has been an increased focus on improving the software process within most organizations. This is in part because of the Year 2000 (Y2K) debacle, the 80–90% failure of large-scale software projects, and a growing realization that following a mature software process is a key determinant to the success of a software project. Starting in the mid-1990s, Rational Corporation was purchasing and merging with other tool companies, and as they did this, they consolidated the processes supported by those tools into a single development process which they named the Unified Process. Is it possible to automate the entire software process? Does Rational have a complete toolset even if it is? I'm not so sure. Luckily, other people were defining software processes — the OPEN Consortium's OPEN process for one and my own process patterns of the Object-Oriented Software Process (OOSP) for another — so we have alternate views of how things should work. These alternate views can be used to drive a more robust view of the Unified Process, resulting in an enhanced lifecycle for the Unified Process that accurately reflects the real-world needs of your organization. I believe that the collected wisdom contained in *Software Development* over the years can be used to flesh-out the Unified Process, truly unifying the best practices in our industry. Hence this book series.

***Following a proven, mature process is key to your success
as a software professional.***

Why is a software process important? Step back for a minute. Pretend you want to have a house built and you ask two contractors for bids. The first one tells you that using a new housing technology, he can build a house for you in two weeks if he starts building first thing tomorrow and it will only cost you $100,000. This contractor has some top-notch carpenters and plumbers that have used this technology to build a garden shed in the past and they're willing to work day and night for you to make this deadline. The second one tells you that she needs to discuss what type of house you would like built, and then once she's comfortable that she understands your needs, she'll put together a set of drawings within a week that you can then review and provide feedback on. This initial phase will cost you $10,000 and once you decide what you want built, she can then put together a detailed plan and cost schedule for the rest of the work. Which contractor are you more comfortable with; the one that wants to start building or the one that wants to first understand what needs to be built, model it, plan it, and then build it? Very obviously, the second contractor has a greater chance at success in delivering a house that meets your actual needs. Now assume that you're having software built — something that is several orders of magnitude more complex and often more expensive than building a house — and assume once again you have two contractors that want to take the exact same approaches. Which contractor are you more comfortable with? I hope the answer is still the second one; the one with a sensible process. Unfortunately, practice shows that the vast majority of the time, organizations appear to choose the approach of the first contractor; that of hacking. Of course, practice also shows that we have roughly an 85% failure rate, do you think the two phenomena are related? I think so.

The Elaboration Phase

The Elaboration phase is the second of five phases — Inception, Elaboration, Construction, Transition, and Production — that a release of software experiences throughout its complete lifecycle. This phase has several goals:

- To produce a proven, architectural baseline for your system
- To evolve your requirements model to the "80% completion point"
- To develop a coarse-grained project plan for the entire Construction phase
- To ensure that the critical tools, processes, standards, and guidelines have been put in place for the Construction phase
- To understand and eliminate the high-priority risks of your project

This book describes a collection of articles written by industry luminaries that describe leading-edge best practices. One goal of this book, and of this series in general, is to provide alternate, proven approaches to the techniques encompassed by the Unified Process. Another goal is to fill the gaps of the Unified Process — to be fair, no process can ever truly be considered "complete." The Unified Process is a development process, not a software process; therefore, just because of its chosen scope, it's going to be missing important concepts for most software professionals. Luckily, the writers appearing in *Software Development* have taken a much broader view of process.

About This Series

This book series is comprised of four volumes: one for the Inception phase, one for the Elaboration phase, one for the Construction phase, and a fourth one for the Transition and Production phases. Each book stands on its own, but for a complete picture of the entire software process, you need to read all four volumes. Because articles are not repeated between volumes, you will find that each one truly does offer you great value. One side effect is that some books are weak in some process workflows where others are very strong. For example, the Inception phase volume contains a wealth of material about project planning and estimating — critical topics at the beginning of a project — resulting in the other three volumes being weak in these topics.

As an aside, I found this series significantly harder to compose than I first thought. The problem that I ran into was that I had a wealth of material to choose from, and only a limited number of pages with which to work. If my editor had allowed it, each of these books would have been close to a thousand pages in length and not around the 304 that they are now. The articles in this series truly are the cream of the crop.

About Me

My favorite topic! An avid reader of *Computer Language* and then *Software Development* for years, I first started writing for the magazine in 1995 and eventually became the object columnist in 1997. I started developing software in the early-1980s, writing code in languages such as Fortran and Basic, and later in the mid-1980s in Turing (don't ask), C, Prolog, and Lisp. In the late 1980s I realized that there was more to life than programming and started picking up skills in user interface design, data modeling, process modeling, and testing while I programmed in COBOL and a couple of fourth-generation languages for IBM mainframes. Disillusioned with structured/procedural techniques, in 1990 I discovered objects and readily jumped into Smalltalk development, then into C++ development, then back to Smalltalk. Having worked at several organizations in mentoring and architectural roles, I decided to combine that experience and apply my skills gained as a teaching assistant at the University of Toronto and get into professional training in the mid-1990s. I quickly learned several things; first, that although I like delivering training courses (and still do so today), I didn't want to do it full time. Second, and more importantly, I learned how to communicate complex concepts in an easy-to-understand manner, such as how to develop object-oriented software. This lead to my first two books — *The Object Primer* (Cambridge University Press, 1995) and *Building Object Applications That Work* (Cambridge University Press, 1997/1998) — which describe the fundamentals of object technology from a developer's point of view. I then decided to follow up with two books that describe the Object-Oriented Software Process (OOSP) in *Process Patterns* (Cambridge University Press, 1998) and *More Process Patterns* (Cambridge University Press, 1999), describing the hard-won experiences that I gained working for one of Canada's leading object technology consulting firms. Since then, I've helped several organizations, large and small, new and established, in a variety of industries to improve their internal software processes, and my latest writing endeavors include this book series as well as co-authoring *The Elements of Java Style* (Cambridge University Press, 2000). I think I've found my niche.

Ahh, book writers. What do they know? —Ben Bova

Chapter 1

Introduction

What is a software process? A software process is a set of project phases, stages, methods, techniques, and practices that people employ to develop and maintain software and its associated artifacts (plans, documents, models, code, test cases, manuals, etc.). Consider the hunting process of Figure 1.1. Although I'm sure the *X-Files* will eventually provide an explanation as to why the Neanderthals became extinct, my theory is that the implementation of an inappropriate process such as the one presented is a likely candidate to explain their downfall. The point is that not only do you need a software process, you need one that is proven to work in practice, a software process tailored to meet your exact needs.

Figure 1.1 **What really happened to the Neanderthals (author unknown).**

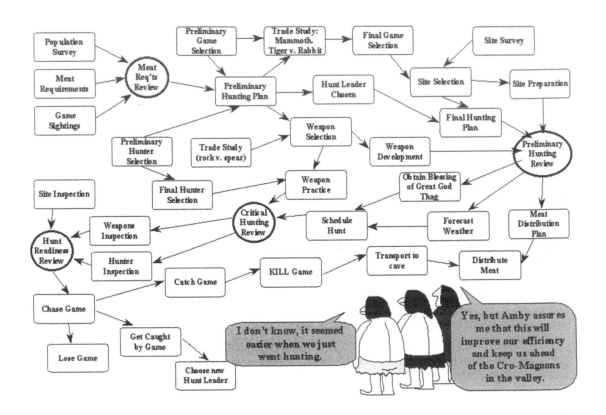

Why do you need a software process? An effective software process will enable your organization to increase its productivity when developing software. First, by understanding the fundamentals of how software is developed you can make intelligent decisions, such as knowing to stay away from SnakeOil v2.0 — the wonder tool that claims to automate fundamental portions of the software process. Second, it enables you to standardize your efforts, promoting reuse and consistency between project teams. Third, it provides an opportunity for you to introduce industry best practices such as code inspections, configuration management, change control, and architectural modeling to your software organization.

An effective software process will also improve your organization's maintenance and support efforts, also referred to as production efforts, in several ways. First, it should define how to manage change and appropriately allocate maintenance changes to future releases of your software, streamlining your change process. Second, it should define, first, how to smoothly transition software into operations and support and then, how the operations and support efforts are actually performed. Without effective operations and support processes, your software will quickly become shelfware.

An effective software process considers the needs of both development and production.

Why adopt an existing software process, or improve your existing process using new techniques? The reality is that software is growing more and more complex, and without an effective way to develop and maintain that software, the chances of you succeeding will only get worse. Not only is software getting more complex, you're also being asked to create more software simultaneously. Most organizations have several software projects currently in development and have many more in production — projects that need to be managed effectively. Furthermore, our industry is in crisis; we're still reeling from the simple transition from using a two-digit year to a four-digit year, a "minor" problem with an estimated price tag of $600 billion worldwide. The nature of the software we're building is also changing — from the simple batch systems of the 1970s for which structured techniques were geared, to the interactive, international, user-friendly, 24/7, high-transaction, high-availability online systems for which object-oriented and component-based techniques are aimed. And while you're doing that, you are asked to increase the quality of the systems that you're delivering, and to reuse as much as possible so that you can work faster and cheaper. A tall order, one that is nearly impossible to fill if you can't organize and manage your staff effectively. A software process provides the basis to help do just that.

Software is becoming more complex, not less.

1.1 The Unified Process

The Unified Process is the latest endeavor of Rational Corporation (Kruchten, 1999), the same people who introduced what has become the industry-standard modeling notation: the Unified Modeling Language (UML). The heart of the Unified Process is the Objectory Process, one of several products and services that Rational acquired when they merged with Ivar Jacobson's Objectory organization several years ago. Rational enhanced Objectory with their own processes (and those of other tool companies that they have either purchased or partnered with) to form the initial version (5.0) of the Unified Process officially released in December of 1998.

Figure 1.2 presents the initial lifecycle of the Unified Process made up of four serial phases and nine core workflows. Along the bottom of the diagram, you see that any given development cycle through the Unified Process should be organized into iterations. The basic concept is that your team works through appropriate workflows in an iterative manner so at the end of each iteration, you produce an internal executable that can be worked with by your user community. This reduces the risk of your project by improving communication between you and your customers. Another risk reduction technique built into the Unified Process is the concept that you should make a "go/no-go" decision at the end of each phase. If a project is going to fail, then you want to stop it as early as possible in its lifecycle — an important concept in an industry with upwards toward an 80–90% failure rate on large-scale, mission-critical projects (Jones, 1996).

Figure 1.2   The initial lifecycle for the Unified Process.

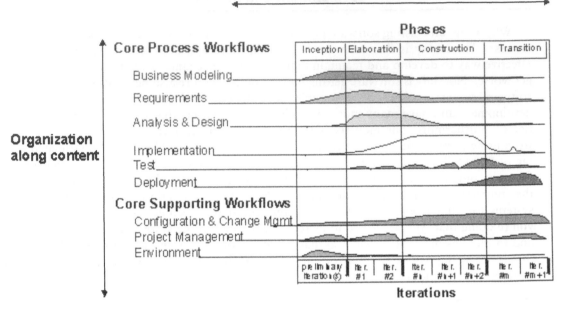

The Inception phase is where you define the project scope and define the business case for the system. The initial use cases for your software are identified and the key ones are described briefly. Use cases are the industry standard technique for defining the functional requirements for systems, providing significant productivity improvements over traditional requirement documents because they focus on what adds value to users as opposed to product features. Basic project management documents are started during the Inception phase, including the initial risk assessment, the estimate, and the project schedule. As you would expect, key tasks during this phase include business modeling and requirements engineering, as well as the initial definition of your environment including tool selection and process tailoring.

You define the project scope and the business case during the Inception phase.

The Elaboration phase, the topic of this volume, focuses on detailed analysis of the problem domain and the definition of an architectural foundation for your project. Because use cases aren't sufficient for defining all requirements, as you'll see in Chapter 4 (which describes the Requirements workflow), a deliverable called a *supplementary specification* is defined which describes all non-functional requirements for your system. A detailed project plan for the Construction phase is also developed during this phase based on the initial management documents started in the Inception phase.

***You define the architectural foundation for your system during
the Elaboration phase.***

The Construction phase, often the heart of software projects (typically to their detriment), is where the detailed design for your application is developed as well as the corresponding source code. I say that projects focus on this phase to their detriment because organizations often do not invest sufficient resources in the previous two phases and therefore lack the foundation from which to successfully develop software that meets the needs of their users. The goal of this phase is to produce the software and supporting documentation to be transitioned to your user base.

You finalize the system to be deployed during the Construction phase.

The purpose of the Transition phase is to deliver the system to your user community. There is often a beta release of the software to your users, typically called a *pilot release* within most businesses, in which a small group of users work with the system before it is released to the general community. Major defects are identified and potentially acted upon during this phase. Finally, an assessment is made regarding the success of your efforts to determine whether another development cycle/increment is needed to further enhance the system.

You deliver the system during the Transition phase.

The Unified Process has several strengths. First, it is based on sound software engineering principles such as taking an iterative, requirement-driven, architecture-based approach to development in which software is released incrementally. Second, it provides several mechanisms, such as a working prototype at the end of each iteration and the "go/no-go" decision point at the end of each phase, which provides management visibility into the development process. Third, Rational has made, and continues to make, a significant investment in their Rational Unified Process product (http://www.rational.com/products/rup), an HTML-based description of the Unified Process that your organization can tailor to meet its exact needs.

The Unified Process also suffers from several weaknesses. First, it is only a development process. The initial version of the Unified Process does not cover the entire software process; as you can see in Figure 1.2, it is very obviously missing the concept of operating and supporting your software once it has been released into production. Second, the Unified Process does not explicitly support multi-project infrastructure development efforts such as organization/enterprise-wide architectural modeling, discussed in Chapter 5 (which describes the proposed Infrastructure Management Workflow), missing opportunities for large-scale reuse within your organization. Third, the iterative nature of the lifecycle is foreign to many experienced developers, making acceptance of it more difficult, and the rendering of the lifecycle in Figure 1.2 certainly doesn't help this issue.

1.2 Moving Beyond the Unified Process

So how do you enhance the Unified Process to make it fit the real-world needs of developers? First, you need to start at the requirements for a process — a good start at which is the Capability Maturity Model (CMM). Second, you should look at the competition, in this case the OPEN Process (Graham, Henderson-Sellers, and Younessi, 1997) and the process patterns of the Object-Oriented Software Process (Ambler 1998b, Ambler 1999), and see which features you can reuse from those processes. Third, you should then formulate an enhanced lifecycle based on what you've learned and support that lifecycle with proven best practices.

1.2.1 The Capability Maturity Model (CMM)

The Software Engineering Institute (SEI) at Carnegie Mellon University (http://www.sei.cmu.edu) has proposed the Capability Maturity Model (CMM) — a framework from which a process for large, complex software efforts can be defined (Software Engineering Institute, 1995). The CMM defines five maturity levels, evolutionary plateaus geared towards achieving a mature software process, that an organization can attain with respect to the software process. According to the CMM, to achieve a specific maturity level, an organization must satisfy and institutionalize all of the key process areas (KPAs) for that level and for the previous levels. Each KPA defines a specific feature, albeit a high-level one, that a software process must exhibit. For example, there are key process areas focusing on quality management, project management, and software engineering. Sponsored by the United States Department of Defense (DoD), the CMM has been adopted by hundreds of organizations worldwide that aim to improve the way that they develop software.

One strength of the CMM is that it provides the basis for the requirements for a mature software process. Organizations with immature software processes are typically reactionary and have little understanding as to how to successfully develop software. Organizations with mature software processes, on the other hand, understand the software process, which enables them to judge the quality of the software products and the process that produces them. Organizations with mature software processes have a higher success rate and a lower overall cost of software across the entire life of a software product than do immature software organizations.

The CMM provides a basis for the requirements for a mature software process.

Why should your organization strive to increase its software process maturity? The answer is two-fold: first, the greater the software process maturity of your organization, the greater the quality of the software products that it produces. The direct result is increased focus on management and project deliverables as the maturity of your organization increases. Second, as the software process maturity of your organization grows, there is a corresponding reduction of risk on your software projects — the result of increased management control and use of measurements (metrics) to understand and manage the software process and the products that you produce.

Software process maturity increases quality and decreases risk.

1.2.2 The OPEN Process

The OPEN consortium, a group of individuals and organizations promoting and enhancing the use of object-oriented technology, has developed a comprehensive software process known as the OPEN Process (http://www.open.org.au) (Graham, Henderson-Sellers, and Younessi, 1997). Like the Unified Process, the OPEN Process is aimed at organizations using object and component technology, although it can easily be applied to other software development technologies. Also similar to the Unified Process, OPEN was initially created by the merger of earlier methods: MOSES, SOMA, Firesmith, Synthesis, BON, and OOram. The OPEN Process supports the UML notation and the Object Modeling Language (OML) notation, and any other OO notation to document the work products the OPEN process produces. The shape of the lines and bubbles may change, but the fundamentals of the software process remain the same.

The OPEN Process is a comprehensive software process that provides significant material beyond the current scope of the Unified Process.

The contract-driven lifecycle for the OPEN process is depicted Figure 1.3. The left-hand side of the figure represents the activities for a single project, whereas the activities in the right-hand side represent cross-project activities. This is a major departure from what you've seen in the Unified Process; the lifecycle of the OPEN Process explicitly includes activities outside of the scope of a single project. This is called *programme management* in the OPEN Process; a programme being a collection of projects and/or releases of an application or suite of applications. Common programme management activities would include organization/enterprise-wide architectural modeling efforts, process improvement efforts, and standards/guidelines development and support. This sort of effort is often called *enterprise management* or *infrastructure management* within software organizations.

Programme management is the management of the portfolio of software projects within your organization.

Figure 1.3   The OPEN contract-driven lifecycle.

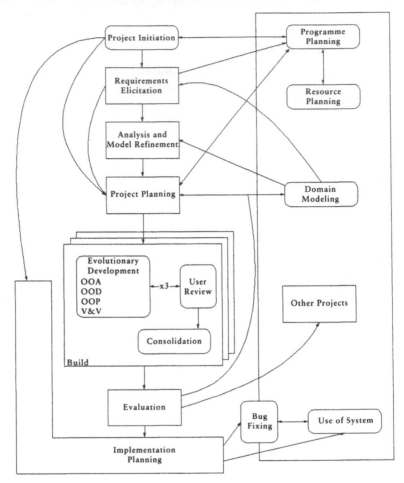

1.2.3 Process Patterns

Similarly, Figure 1.4 depicts the lifecycle of the Object-Oriented Software Process (OOSP), comprised of a collection of process patterns (Ambler 1998; Ambler, 1999). Similar to the Unified Process, the OOSP is based on the concept that large-scale, mission-critical software development is fundamentally serial in the large, iterative in the small, delivering incremental releases in Internet time. A *process pattern* is a collection of general techniques, actions, and/or tasks (activities) that solve a specific software process problem taking the relevant forces/factors into account. Just like design patterns describe proven solutions to common software design problems, process patterns present proven solutions to common software process patterns. Process patterns were originally proposed by James Coplien in his paper, "A Generative Development-Process Pattern Language" in the first *Pattern Languages of Program Design* book (Addison-Wesley, 1995). I believe that there are three scales of process patterns: phase process patterns, stage process patterns, and task process patterns. A *phase process pattern* depicts the interactions between the stage process patterns for a single project phase, such as the Initiate and Deliver phases of the OOSP. A *stage process pattern* depicts the tasks, which are often performed iteratively, of a single project stage such as the Program the Model stage. Finally, *task process patterns* address lower-level process issues, such as the Reuse First process pattern that describes how to achieve significant levels of reuse within your organization and the Technical Review process pattern which describes how to organize reviews and inspections. The three scales of process patterns are conceptually similar to the scale of other types of patterns — in the modeling world you have architectural patterns, design patterns, and programming idioms.

Process patterns are reusable building blocks from which you can tailor a software process to meet your organization's unique needs.

Figure 1.4   The Object-Oriented Software Process (OOSP) lifecycle.

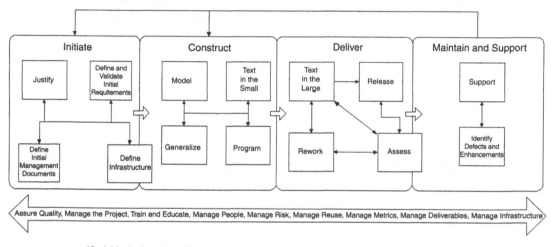

"Serial in the large, iterative in the small, delivering incremental releases over time."

In Figure 1.4 you see that there are four project phases within the OOSP — Initiate, Construct, Deliver, and Maintain and Support — each of which is described by a corresponding phase process pattern. There are also fourteen project stages in the OOSP — Justify, Define and Validate Initial Requirements, Define Initial Management Documents, Define Infrastructure, Model, Program, Test In The Small, Generalize, Test In The Large, Rework, Release, Assess, Support, and Identify Defects and Enhancements — each of which is described by a stage process pattern. Project stages are performed in an iterative manner within the scope of a single project phase. Project phases, on the other hand, are performed in a serial manner within the OOSP. This is conceptually very similar to what you see in the Unified Process (Figure 1.2) — comprised of four serial phases and nine workflows that are performed in an iterative manner.

The OOSP includes production issues such as the maintenance and support of your software.

As you can see, there is more to the OOSP than just its phases and stages. The "big arrow" at the bottom of the diagram indicates important tasks critical to the success of a project — tasks that are applicable to all stages of development. These tasks include quality assurance, project management, training and education, people management, risk management, reuse management, metrics management, deliverables management, and infrastructure management. The important thing to note is that several of these tasks are applicable at both the project and cross-project (programme) levels. For example, risk management should be performed for both a single project and for your portfolio of software projects. It's no good if each project is relatively risk free if, as a collection, they become quite risky. Deliverables management, which includes configuration management and change control functions, is also applicable for both a single project and for a collection of projects because the collection of projects may share a single artifact, such as common source code. Infrastructure management — where you evolve your corporate processes, standards, guidelines, and architectures — is obviously applicable across several projects but will also apply to a single project because that project team will need guidance and support to use your organization's software infrastructure. More information regarding process patterns can be found at The Process Patterns Resource Page, http://www.ambysoft.com/processPatternsPage.html.

The OOSP includes both project and cross-project activities.

1.3 The Enhanced Lifecycle For the Unified Process

You've seen overviews of the requirements for a mature software process and the two competing visions for a software process, so knowing this, how do you complete the Unified Process? Well, the first place to start is to redefine the scope of the Unified Process to include the entire software process, not just the development process. This implies that processes for operations, support, and maintenance efforts need to be added to the Unified Process. Second, to be sufficient for today's organizations, the Unified Process also needs to support the management of a portfolio of projects, something the OPEN Process has called "programme management" and the OOSP has called "infrastructure management." These first two steps result in the enhanced version of the lifecycle depicted in Figure 1.5. Finally, the Unified Process needs to be fleshed out with proven best practices; in this case, found in articles published in *Software Development*.

Figure 1.5   The enhanced lifecycle for the Unified Process.

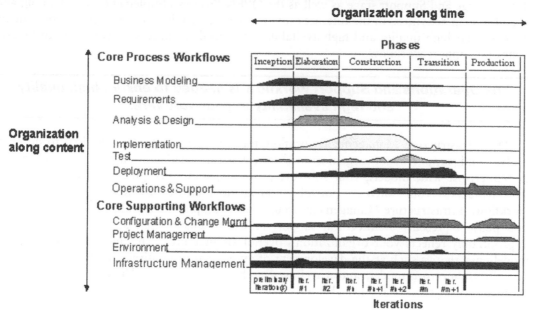

The enhanced lifecycle includes a fifth phase, Production, representing the portion of the software lifecycle after a system has been deployed. As the name of the phase implies, its purpose is to keep your software in production until it is either replaced with a new version — from a minor release such as a bug fix to a major new release — or it is retired and removed from production. Note that there are no iterations during this phase (or there is only one iteration depending on how you wish to look at it) because this phase applies to the lifetime of a single release of your software. To develop and deploy a new release of your software, you need to run through the four development phases again.

The Production phase encompasses the post-deployment portion of the lifecycle.

Figure 1.5 also shows that there are two new workflows: a process workflow called *Operations & Support* and a supporting workflow called *Infrastructure Management*. The purpose of the Operations & Support workflow is exactly as the name implies: to operate and support your software. Operations and support are both complex endeavors, endeavors that need processes defined for them. This workflow, as do all the others, spans several phases. During the Construction phase, you will need to develop operations and support plans, documents, and training manuals. During the Transition phase, you will continue to develop these artifacts, reworking them based on the results of testing and you will train your operations and support staff to effectively work with your software. Finally, during the Production phase, your operations staff will keep your software running, performing necessary backups and batch jobs as needed, and your support staff will interact with your user community in working with your software. This workflow basically encompasses portions of the OOSP's Release stage and Support stage as well as the OPEN Process's Implementation Planning and Use of System activities. In the Internet economy, where you do 24/7 operations, you quickly discover that high quality and high availability is crucial to success — you need an Operations and Support workflow.

The Operations and Support workflow is needed to ensure high quality and high availability in your software.

The Infrastructure Management workflow focuses on the activities required to develop, evolve, and support your organization's infrastructure artifacts such as your organization/enterprise-wide models, your software processes, standards, guidelines, and your reusable artifacts. Your software portfolio management efforts are also performed in this workflow. Infrastructure Management occurs during all phases; the blip during the Elaboration phase represents architectural support efforts to ensure that a project's architecture appropriately reflects your organization's overall architecture. This includes infrastructure modeling activities such as the development of an enterprise requirements/business model, a domain architecture model, and a technical architecture model. These three core models form the infrastructure models that describe your organization's long-term software goals and shared/reusable infrastructure. The processes followed by your Software Engineering Process Group (SEPG) — responsible for supporting and evolving your software processes, standards, and guidelines — are also included in this workflow. Your reuse processes are too; practice shows that to be effective, reuse management is a cross-project endeavor. For you to achieve economies of scale developing software, increase the consistency and quality of the software that you develop, and increase reuse between projects, you need to manage your common infrastructure effectively. You need the Infrastructure Management workflow.

Infrastructure Management supports your cross-project/programme-level, activities such as reuse management and organization/enterprise-wide architecture.

Comparing the enhanced lifecycle (Figure 1.5) with the initial lifecycle (Figure 1.2), you will notice that several of the existing workflows have also been updated. First, the Test workflow has been expanded to include activity during the Inception phase. You develop your initial, high-level requirements during this phase — requirements that you can validate using techniques such as walkthroughs, inspections, and scenario testing. Two of the underlying philosophies of the OOSP are that a) you should test often and early and, b) that if something is worth developing, then it is worth testing. Therefore testing should be moved forward in the lifecycle. Also, the Test workflow also needs to be enhanced with the techniques of the OOSP's Test In The Small and Test In The Large stages.

Test early and test often. If it is work creating, it is worth testing.

The second modification is to the Deployment workflow — extending it into the Inception and Elaboration phases. This modification reflects the fact that deployment, at least of business applications, is a daunting task. Data conversion efforts of legacy data sources are often a project in their own right, a task that requires significant planning, analysis, and work to accomplish. Furthermore, my belief is that deployment modeling should be part of the Deployment workflow, and not the Analysis & Design workflow as it currently is, due to the fact that deployment modeling and deployment planning go hand-in-hand. Deployment planning can and should start as early as the Inception phase and continue into the Elaboration and Construction phases in parallel with deployment modeling.

Deployment is complex and planning often must start early in development to be successful.

The Environment workflow has been updated to include the work necessary to define the Production environment, work that would typically occur during the Transition phase. The existing Environment workflow processes effectively remain the same, the only difference being that they now need to expand their scope from being focused simply on a development environment to also include operations and support environments. Your operations and support staff need their own processes, standards, guidelines, and tools, the same as your developers. Therefore, you may have some tailoring, developing, or purchasing to perform to reflect this need.

The Configuration & Change Management workflow is extended into the new Production phase to include the change control processes needed to assess the impact of a proposed change to your deployed software and to allocate that change to a future release of your system. This change control process is often more formal during this phase than what you do during development due to the increased effort required to update and re-release existing software. Similarly, the Project Management workflow is also extended into the new Production phase to include the processes needed to manage your software once it has been released.

Change control management will occur during the Production phase.

The Project Management workflow is expanded in the enhanced lifecycle for the Unified Process. It is light on metrics management activities and subcontractor management — a CMM level 2 key process area — a key need of any organization that outsources portions of its development activities or hires consultants and contractors. People management issues, including training and education as well as career management, are barely covered by the Unified Process because those issues were scoped out of it. There is far more to project management than the technical tasks of creating and evolving project plans; you also need to manage your staff and mediate the interactions between them and other people.

There is far more to project management than planning, estimating, and scheduling.

1.4 The Goals of the Elaboration Phase

During the Elaboration phase, your project team will focus on evolving the requirements for your application and on the identification, development, and verification of your project's architecture. Your team will work towards the following goals:

- Establish a solid understanding of the most critical requirements
- Evolve the requirements model to the "80% point"
- Define and validate an architectural baseline for your project
- Eliminate high-priority risks via modeling and prototyping
- Baseline the vision for your project
- Demonstrate, to all project stakeholders, that your architecture supports your project vision
- Develop and baseline a high-fidelity plan for the Construction phase
- Define the project team's environment, including the tools, processes, and guidelines that you will employ

As you work towards your goals, you will create and/or evolve a wide variety of artifacts:

- Requirements model (use cases, supplementary specifications) describing what you intend to build
- Project-level software architecture document (SAD) describing how you intend to build it
- Executable architectural prototype showing that your architecture works
- Revised project schedule, estimate, and risk list to manage your project
- Revised business case justifying your project
- Revised team environment definition (tools, processes, guidelines, standards, etc.) indicating the tools and techniques that your team will use
- Preliminary user, support, and operations documentation that define how your customers will use your system

1.5 How Work Generally Proceeds During the Elaboration Phase

A fundamental precept of the Unified Process is that work proceeds in an iterative manner throughout the activities of the various workflows. However, at the beginning of each iteration you will spend more time in requirements-oriented activities and towards the end of the iteration your focus will be on test-oriented activities. As a result, to make this book easier to follow, the chapters are organized in the general order by which you would proceed through a single iteration of the Elaboration phase. Figure 1.5 indicates the workflows applicable during the Elaboration phase:

- Project Management (Chapter 2)
- Business Modeling (Chapter 3)
- Requirements (Chapter 4)
- Infrastructure Management (Chapter 5)
- Analysis & Design (Chapter 6)
- Implementation (covered in detail in Vol. 3: *The Unified Process Construction Phase*)

- Test (Chapter 7)
- Deployment (covered in detail in Vol. 4: *The Unified Process Transition Phase*)
- Configuration & Change Management (covered in detail in Vol. 3: *The Unified Process Construction Phase*)
- Environment (covered in detail in Vol. 1: *The Unifed Process Inception Phase*)

1.5.1 The Project Management Workflow

The purpose of the Project Management workflow is to ensure the successful management of your project team's efforts. During the Elaboration phase, the Project Management workflow includes several key activities:

Manage risk. The project manager must ensure that the project's risks are managed effectively, and the best way to do that is to ensure that the high-risk use cases are covered by the technical prototyping efforts of this phase.

Manage the technical prototyping efforts. You must also ensure that the project team stresses the chosen technology, thereby proving that your chosen approach works. This helps to reduce the risk to your project as well as to thwart political roadblocks such as people claiming that "it will never work."

Plan the project. Part of project management is the definition of a detailed plan for this phase — it is common to have between one and three iterations during the Elaboration phase, and to devise a coarse-grained plan for the Construction phase. When defining iterations, a smaller iteration allows you to show progress and thereby garner support for your efforts, whereas bigger iterations are typically an indication that your project is in trouble and has fallen into a serial mindset. Your planning efforts will include the definition of an estimate and a schedule.

Define and build of your project team for the Construction phase. This effort should not be underestimated as it is very difficult to find a team of people with the necessary skills that can work together well.

Navigate the political waters within your organization. Although politics appear to be beyond the scope of the initial version of the Project Management workflow (Kruchten, 1999), the fact remains that softer issues such as people management and politics are a reality for all software projects.

Achieve architectural consensus. There must be consensus regarding the project's architecture both within the project team and within your organization for your team to be successful. You should not move into the Construction phase without such consensus, otherwise your team will either start fighting within itself over what is "the best architecture" or will constantly be bombarded by outsiders attacking them with "better approaches."

Measure the progress of your development efforts. You need to track and report your status to senior management, and to the rest of your organization. This requires you to record key measurements, called *metrics*, such as the number of work days spent and the number of calendar days taken.

Define and/or manage relationships with subcontractors and vendors. It is common to see part or all of the work of a project outsourced to an information technology company that specializes in that kind of work. However, you still need to manage the company that you have outsourced the work to.

Assess the viability of the project. At the end of the Elaboration phase, and perhaps part way through it, a project viability assessment should be made (also called a "go/no-go" decision) regarding the project. Remember, 85% of all large-scale projects fail so you are doing your organization a favor by stopping a project early in its lifecycle when it looks like it is doomed.

There is more to project management than planning, estimating, and scheduling.

1.5.2 The Business Modeling Workflow

The purpose of the Business Modeling workflow is to model the business context of your system. During the Elaboration phase, the Business Modeling workflow includes several key activities:

Development of a context model. A context model shows how your system fits into its overall environment. This model will depict the key organizational units that will work with your system, perhaps the marketing and accounting departments, and the external systems that it will interact with.

Develop a business requirements model. This model contains a high-level use case model, typically a portion of your enterprise model (perhaps modeled in slightly greater detail), that shows what behaviors you intend to support with your system. The model also includes a glossary of key terms and optionally, a high-level class diagram (often called a *Context Object Model*) that models the key business entities and the relationships between them. The business model has an important input for your Requirements workflow efforts.

Evolve a common understanding of the system with stakeholders. You need to reach consensus between your project stakeholders as to what your project team will deliver. Stakeholders include: your direct users, senior management, user management, your project team, your organization's architecture team, and potentially even your operations and support management. Without this common understanding, your project will likely be plagued with politics and infighting and could be cancelled prematurely if senior management loses faith in it.

Develop a business process model. A business process model, traditionally called an analysis data-flow diagram in structured methodologies, depicts the main business processes, the roles/organizations involved in those processes, and the data flow between them. Business process models show how things get done, as opposed to a use case model that shows what should be done.

Your business model shows how your system fits into its environment and helps you to evolve a common understanding with your project stakeholders.

1.5.3 The Requirements Workflow

The purpose of the Requirements workflow is to engineer the requirements for your project. During the Elaboration phase, the Requirements workflow includes several key activities:

Evolve the requirements model to the "80% level".   The basic idea is that your goal is to understand the core 80% of the system that you are building — you might be missing a few details, but you understand what it is that you are trying to build. A good way to know that you are 80% of the way there is that your requirement gathering efforts have shifted from asking fundamental questions about the problem domain to focus almost solely on details.

Evolve all aspects of the requirements model.   The requirements model is composed of far more than just use cases; you have a glossary defining key business terms, a user interface prototype, and a Supplementary Specifications document that defines technical, performance, system, and other various non-behavioral requirements. All aspects of the requirements model must be evolved, not just the use cases.

Ensure that the system may be delimited.   You need to define the scope of what it is that you will build, as well as what you will not build. Requirements are a primary input into the scoping of your system — a key activity of the Business Modeling workflow and then protected by your Configuration and Change Management workflow. To do this. you will start assigning requirements to releases of your system, and for requirements within the release that you are currently working on, you will begin assigning those to phases and perhaps even iterations.

Define sufficient information to plan the Construction phase.   Your   requirements model defines what you intend to build, therefore to plan your construction efforts, you need to define your requirements so that you have enough information from which to plan.

Prototype the user interface for the system.   User interface prototyping is a key requirements engineering activity, and will be an important part of your requirements effort during this phase. The user interface is the system for your users — this is the part that you want to get right if you intend to keep them happy.

Ensure that the requirements are testable.   Your requirements model, and not just your use case model, is a primary input into the definition of your test plan and your test cases. Your requirements analysts will need to work together will your test analysts to ensure that what they are defining is understandable and sufficient for the needs of testing.

Use cases are only a start at the requirements for your system.

1.5.4 The Infrastructure Management Workflow

The Infrastructure Management workflow encompasses activities that are outside of the scope of a single project, yet are still vital to your organization's success. During the Elaboration phase, the Infrastructure Management workflow includes the following key activities:

Manage and support reuse.   Strategic reuse management is a complex endeavor, one that spans all of the projects within your organization. Your team should strive to identify and reuse existing artifacts wherever possible, to buy instead of build where that makes sense, and to produce high-quality artifacts that can potentially be generalized for reuse.

Perform programme management. Programme management is the act of managing your organization's portfolio of software projects — projects that are either in development, in production, or waiting to be initiated. During the Elaboration phase, you will need to ensure that your project fits into the overall picture; few projects are islands unto themselves.

Perform enterprise requirements modeling. Enterprise requirements modeling (Jacobson, Griss, Jonsson, 1997) is the act of creating a requirements model that reflects the high-level requirements of your organization. Your project's requirements model should reflect, in great detail, a small portion of this overall model. As you evolve your requirements model, you will need to ensure that the two models are consistent with one another.

Perform organization/enterprise-level architectural modeling. Although your individual system may have its own unique architecture, it needs to fit in with your organization's overall business/domain and technical architecture. Your project's architecture should start with the existing architecture as a base and then ensure that any deviations fit into the overall picture.

Perform organization/enterprise-wide process management. Your project team may discover, through its Environment workflow efforts, that existing corporate processes, standards, and guidelines need to be evolved to meet the new needs of the business.

Strategic reuse management, enterprise requirements modeling, organization/enterprise-wide architecture, and process management are infrastructure issues that are beyond the scope of a single project.

1.5.5 The Analysis and Design Workflow

The purpose of the Analysis and Design workflow is to model your software. During the Elaboration phase, the Analysis and Design workflow includes the following key activities:

Develop a design model based on your requirements. The primary input into this workflow is your requirements model, from which you perform modeling techniques such as class modeling, sequence diagramming, collaboration diagramming, state modeling, and component modeling. The end result will be a design model for your system.

Define and evolve a robust architecture for your system. Practice shows that one of the most effective ways to develop software is to define the architecture for your system first, then to perform detailed modeling and implementation based on that architecture. Your project's architecture must reflect *both* the requirements for your system as well as the existing architecture of your organization. During the Elaboration phase, architecture-level modeling will be the primary focus of your analysis and design efforts.

Adapt the design to your implementation environment. Your design must not only reflect your requirements, it should also reflect the target environment of your organization. For example, a design meant to be deployed into a highly distributed environment may not be appropriate for an environment of stand-alone, disconnected personal computers.

Finalize the user interface design.   Although user interface prototyping is an important activity of the Requirements workflow, the user interface design effort is actually part of this workflow. The purpose of user interface prototyping is to understand the requirements for your software and to communicate your understanding of those requirements. The prototype is then evolved to conform to your organization's accepted user interface design standards.

Your key goal is to develop the architecture for your system.

1.5.6 The Implementation Workflow

The purpose of the Implementation workflow is to write and initially test your software. During the Elaboration phase, the Implementation workflow includes the following key activities:

Document code.   Code that isn't worth documenting isn't worth writing. Furthermore, practice shows that developers who write the initial documentation for their code, even if it's only in abbreviated form, are significantly more productive than those that don't. The lesson is simple: think first, then act. However, documentation isn't as important during the Elaboration phase because the majority of the source code that you write will be for your technical prototype and/or for the user interface prototype of the system. This code has a tendency to be sloppily written because it is going to be thrown away, or at least it should be, therefore there isn't as great of a need to document this code.

Write code.   If you can't reuse something that already exists, then you're unfortunately forced to write new source code. Your new source code will be based on the design model, and when you find issues with the design model (nothing is ever perfect) you need to work closely with the modelers to address the issues appropriately. Sometimes the model will need to change, sometimes your source code will.

Test code.   There is a multitude of testing techniques, such as coverage testing, white-box testing, inheritance-regression testing, class testing, method testing, and class-integration testing (to name a few) that you will use to validate your code.

Integrate code.   You need to integrate the work of your entire team, ideally in regular intervals, so that it may all work together. The end result of your integration efforts should be a working build of your system.

1.5.7 The Deployment Workflow

The purpose of the Deployment workflow is to ensure the successful deployment of your system. During the Elaboration phase, the Deployment workflow includes the following key activities:

Plan your deployment efforts.   Deployment of software, particularly software that replaces or integrates with existing legacy software, is a complex task that needs to be thoroughly planned. Your deployment plan may have been started during the Inception phase (see Vol. 1 in this series), and if not, it likely will be started during Elaboration.

Create a deployment model.   Originally an activity of the Analysis and Design workflow in the initial version of the Unified Process (Kruchten, 1999), this work has been moved into the Deployment workflow for the enhanced lifecycle. The reason for this is simple: deployment planning and deployment modeling go hand-in-hand and are the main drivers of your actual deployment efforts.

Make inroads with operations and support departments. You need the support of your operations and support departments to successfully deploy your software, and the earlier that you start working together with them, the greater the chance of your project being accepted by them.

1.5.8 The Test Workflow

The purpose of the Test workflow is to verify and validate the quality and correctness of your system. During the Elaboration phase, the Test workflow includes the following key activities:

Stress test the technical prototype. A main goal of the Elaboration phase is to define and prove the architectural baseline for your system, and a major part of proving it is to stress test it. The goal of stress testing is to determine where something breaks — perhaps your architecture can only handle 500 simultaneous users producing no more that 1,500 transactions per second.

Inspect your requirements, design, and/or implementation models. If you can build it, you can test it, and anything that isn't worth testing, likely isn't worth building. It is possible to test your requirements model; you can do a user interface walkthrough, a use-case model walkthrough, or even use-case scenario testing (Ambler, 1998a). It is possible to test your design; you can perform peer reviews and inspections, and the same thing can be said about your implementation model.

Provide input into the project viability assessment. The purpose of the project viability assessment (a key activity of the Project Management workflow) is to determine whether or not it makes sense to continue working on your project. Important items of information that are input into this decision are whether or not the architecture will work in a production environment and whether or not the requirements accurately reflect the needs of your users. This information is gathered as part of your testing efforts.

1.5.9 The Configuration and Change Management Workflow

The purpose of the Configuration and Change Management workflow is to ensure the successful deployment of your system. During the Elaboration phase, the Configuration and Change Management workflow includes the following key activities:

Place project artifacts under configuration management control. Configuration management (CM) is essential to ensure the consistency and quality of the numerous artifacts produced by the people working on a project — helping to avoid confusion amongst team members. Without CM, your team is at risk of overwriting each other's work, potentially losing significant time to replace any lost updates. Furthermore, good CM systems will notify interested parties when a change has been made to an existing artifact, allowing them to obtain an updated copy if needed. Finally, CM allows your team to maintain multiple versions of the same artifact, allowing you to rollback to previous versions if need be.

Prioritize and allocate requirements changes. You need to manage change or it will manage you. As your project moves along, new requirements will be identified and existing requirements will be updated and potentially removed. These changes need to be prioritized and allocated to the appropriate release, phase, and/or iteration of your project. Without change control, your project will be subject to what is known as *scope creep* — the addition of requirements that were not originally agreed to be implemented.

1.5.10 Environment Workflow

The purpose of the Environment workflow is to configure the processes, tools, standards, and guidelines to be used by your project team. During the Elaboration phase, the Environment workflow includes the following two key activities:

Tailoring your software process. Although the definition and support of your organization's software process is an activity of the Infrastructure Management workflow (Chapter 5), you still need to tailor that process to meet your project's unique needs. This effort includes the selection and tailoring of software processes, standards, and guidelines. The end result of this effort is called a *development case* (don't ask me why). This effort should have started in the Inception phase but typically continues into Elaboration. This activity is covered in detail in Vol. 1 in this series — which describes best practices for the Inception phase.

Tool selection. Software process, architecture, organizational culture, and tools go hand-in-hand. If they are not already imposed upon your team by senior management, you need to select a collection of development tools such as a configuration management system, a modeling tool, and an integrated development environment. Your tools should work together and should reflect the processes that your team will follow.

1.6   The Organization of This Book

This book is organized in a simple manner. There is one chapter for each workflow of the Elaboration phase, with the exception of the deployment workflow which is covered in detail in the third volume in this series, *The Unified Process Construction Phase*. Each chapter is also organized in a straightforward manner, starting with my thoughts about best practices for the workflow, followed by comments about the *Software Development* articles that I have chosen for the workflow, then ending with the articles themselves. The workflow chapters are then followed by a short chapter wrapping up the book and providing insights into the next phase of the enhanced lifecycle for the Unified Process — the Construction phase.

Chapter 2

Best Practices for the Project Management Workflow

When you are following the enhanced lifecycle of the Unified Process, your Project Management workflow efforts during the Elaboration phase will focus on:

- The technical aspects of project management, such as planning, estimating, scheduling, risk management, and metrics management
- The people aspects of project management such as building your team and managing the people that work on your team
- The political aspects of project management such as coordinating your efforts with other teams
- The training and education efforts required by your staff.

This philosophy, depicted in Figure 2.1, differs from that of the Rational Unified Process (RUP) which focuses mainly on the technical aspects of project management (Kruchten, 1999). Yes, project planning and risk management are very important but the softer aspects of project management are just as important, likely more so, and deserve equal time in your organization's software process.

Figure 2.1   Project management in the real world.

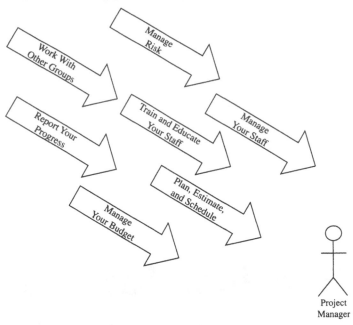

During the Elaboration phase, the Project Management workflow concentrates on achieving several key goals:

- planning in detail and then managing the iteration(s) of the phase,
- developing a coarse-grain plan and estimate for the Construction phase,
- defining and then building your project team for the Construction phase,
- achieving consensus within your organization as to the architecture of the system that you are building,
- and measuring the progress of your development efforts.

The technical aspects of project management, such as planning and estimating, are covered in detail in volume I, *The Unified Process Inception Phase*, of this series. However, these are still several issues that need to be addressed in detail. To complete the Project Management workflow for the Elaboration phase, you need to include the following topics:

- Managing Multi-Team Development
- Managing People
- Managing the Recruitment and Team-Definition Process
- Managing Your Training and Education Efforts

2.1 Managing Multi-Team Development

Most large-scale, mission-critical software projects are developed by several teams of professionals working in concert. The manager of these teams needs to understand the added level of complexity that this entails to be successful. Team members also need to understand the special issues involved with multi-team development to be an effective member of the overall effort.

Large systems are often developed by several teams in concert.

It is common for the work of large projects to be distributed among several development teams working in parallel, often teams that are geographically dispersed. Having been involved with a project that performed work in Anchorage, Toronto, and Paris, I know how difficult it can be to manage global teams. There are several advantages to taking a global development approach, including the ability to develop software twenty-four hours a day (instead of the usual eight to twelve that a single team can accomplish) and the increase in quality resulting from having several sets of eyes looking at your work while it is being developed. These advantages come at a price: the increased overhead of organizing and managing these teams. In section 2.5.1 "Managing Global Teams" (*Software Development*, August 1998), Johanna Rothman provides strategies and techniques for planning, organizing, and then managing dispersed development teams. Your Elaboration efforts may not be dispersed, in fact, you typically want to centralize your initial architectural development efforts to ensure consistency. But if you intend to perform the Construction using dispersed teams, then you need to start planning for that now.

Know thy enemy (it's likely you).

The success of a team is often determined by its ability to collaborate with people in other teams and/or organizations, and in section 2.5.2 "Managing Collaborations" (*Software Development*, January 1998), Mary Loomis provides several insights into successful collaboration. The strategies that she describes are applicable to teams that need to interact with other teams within their organization effectively, with a corporate politics situation, and/or with teams that need to collaborate with teams in other organizations or corporate divisions. You need to plan and to organize your effort as well as manage the collaborations *while* they occur as well as afterwards.

2.2 Managing People

There is far more to project management than creating schedules and plans — you also need to find the right people for your team, and then nurture that team and the individuals on it. In section 2.5.3 "Implementing Feature Teams" (*Software Development*, December 1995), Jim McCarthy provides best practices for managers trying to build an interdisciplinary team whose goal is to develop and deliver a specific feature of product. That feature may be the help system for your application, the spell-checker of a word processor, the reporting subsystem for a large application, or the main editing screens of an application. A foundational concept of the Unified Process is to have multidisciplinary teams, and McCarthy provides excellent advice for building and managing such teams effectively.

Project management is far more than the creation and update of management documents.

2.3   Managing the Recruitment and Team-Definition Process

During the Elaboration phase, you will continue the difficult task of building your development team, sometimes ramping up from a team of several people at the start of the phase to several hundred at the end of it. Teams are built from people currently within your organization as well as new hires. Regardless, you will find that you need to interview and recruit new members to your project team. However, you don't want to overstaff your project early in the lifecycle as this often results in serious trouble for the project team. It is common that project managers will often be given more programmers than they actually need at the beginning of a project, typically during the Elaboration phase, but sometimes even as early as the Inception phase. This often results in the manager cutting key activities such as architectural prototyping and modeling in favor of programming to make everyone look busy. The point to be made is that you want to strive to have the right people at the right time for your project.

In section 2.5.4 "Tips for Technical Interviewing" (*Software Development*, September 1996), Barbara Hanscome presents excellent advice for interviewing potential candidates. It is very difficult to hold a technical interview effectively, and the tips and techniques presented by Hanscome will dramatically increase your chance of success. Simple advice, such as asking a programmer for a sample of their code or a modeler to draw a design for a defined problem, enables you to judge the quality of their work — a key goal of a technical interview. Hanscome points out that you want to determine *how* a person learns as well as how quickly — an important qualification in an industry with rapidly changing technology. Furthermore, the article provides insights into determining how well they will fit on your team and how well they understand the impact of their work. Considering that software spends 80% of its life in maintenance, it's important that your team members understand how the decisions they make during development will affect the maintainability and extensibility of their work

---

Technical interviewing best practices help you to build better project teams.

2.4   Managing Your Training and Education Efforts

In the early 1990s, I spent a lot of time mentoring and training people in object technology, and in fact, I still spend a portion of my time doing so. One of the realizations that I've come to is that any given skill has an expected period of time for which it is in demand (see Figure 2.2). My experience is that programming skills, such as C++ or Java programming, will be in high demand over a period of three to five years and then will dwindle in popularity. These two languages are, in fact, perfect examples — in the mid-1990s, C++ programmers where the most sought after developers. Five years later, as I am writing this, the most sought after people are programmers with Enterprise JavaBean (EJB) experience. Five years from now I expect that another language will have replaced Java. Similarly, modeling skills, such as use case development or object-oriented design, have an expected lifespan of fifteen to twenty years because they have a broader scope of applicability to software development. The skills with the greatest longevity in the information technology (IT) industry, it appears, are project management and people/communication skills — having an expected lifespan of thirty to forty years.

Skills have differing expected lifespans.

Figure 2.2   Comparing the expected lifespan of skills.

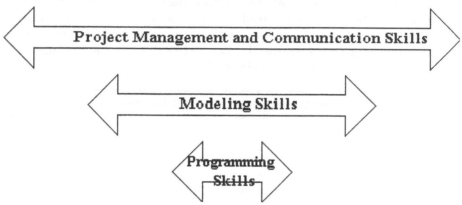

There are three key implications to Figure 2.2:

1. The more technical a skill, the shorter the period of its usefulness to you. This should not be a surprise — the rapid rate of change in the technology field necessitates an equally rapid change in the skills of technologists.

2. If you choose to specialize in a very technical skill (and many developers do), then you need to be prepared to overhaul your skillset regularly. Yes, you might be able to stay with a single language for fifteen or twenty years — consider COBOL and C programmers — but then again look at the sparse employment opportunities for APL, Eiffel, and FOR-TRAN programmers.

3. If you choose to stick to skills with greater longevity, for example project management, you still need to be prepared to keep current with new technologies. In many ways, project managers have an even greater need than programmers for training — programmers specializing in a single language need to gain detailed skills in a reasonably narrow topic, whereas project managers need to gain a shallow knowledge of a broad range of skills.

Understanding when and why to retrain is critical to your career success.

In section 2.5.5 "Constant, Unceasing Improvement" (*Software Development*, January 1995), an excerpt of his book *Debugging the Development Process* (1994), Steve Maguire argues that you should consider the needs of developers to help expand their skills on your projects. An important aspect of people management is that it is both a project and a cross-project activity — an aspect that Maguire brings home with his point that developers need to widen their skills to be successful over time. Developers that specialize in a narrow technology, such as being the embedded Java development expert for Acme-brand widgets, often find themselves in trouble when the market shifts from that technology — a key lesson implied in Figure 2.2. Maguire argues that project managers are responsible for assigning tasks to developers that will provide opportunities to grow and enhance their skills — that assigning developers the same type of tasks over and over again (because they are an expert at that task) is neither good for them nor for your organization. He also suggests that you avoid the "New Year's Syndrome" — the tendency of organizations to focus all of their people management efforts on annual reviews — and instead, include these efforts as part of your regular project management activities. Think of it like this: are you managing a project, are you managing people, or are you managing both?

The needs of the developer outweigh the needs of the project.

In section 2.5.6 "Object-Oriented Training" (*Software Development*, September 1996), I present a collection of best practices for training your developers in object-development techniques, ranging from traditional training courses to bag-lunch sessions to learning teams. Training strategies for key project roles are presented; you need to take a different approach to training programmers than you will for training senior executives. A fundamental concept discussed in the article is that you need to follow-up initial training with mentoring, refresher courses, and written support materials such as books and magazines — a person doesn't become a software engineer just by taking a two-day training course. Tips for overcoming common challenges such as the "I've been doing this all along syndrome" and the fact that not everyone learns the same way are also presented.

Training is just a start — obtain mentoring, build a library of key books, and purchase subscriptions to key magazines and journals.

2.5   The Articles

2.5.1 "Managing Global Teams" by Johanna Rothman, August 1998

2.5.2 "Managing Collaborations" by Mary Loomis, January 1998

2.5.3 "Implementing Feature Teams" by Jim McCarthy, December 1995

2.5.4 "Tips for Technical Interviewing" by Barbara Hanscome, September 1996

2.5.5 "Constant, Unceasing Improvement" by Steve Maguire, January 1995

2.5.6 "Object-Oriented Training" by Scott W. Ambler, September 1996

2.5.1 "Managing Global Teams"
by Johanna Rothman, August 1998

With the added stresses of differing cultures, languages, and time zones — managing a global team requires specific considerations. Here are six tips for making your global software development efforts work.

More companies are looking toward globally dispersed software development teams to solve project staffing problems and make critical time-to-market deadlines. This trend is a fundamental change in how software projects are organized and implemented. Using the idea of "concurrent engineering" to deliver projects faster, you break up a project into smaller, less complex pieces and hire staff scattered throughout the globe who work asynchronously around-the-clock. (This is not the same as adding more people to a project to make it later, à la Brooks's Law.)

Although companies using globally dispersed teams can tap into skilled (and sometimes less costly) developers throughout the world, these teams typically suffer greater communication problems during software development. These issues come about primarily because of culture, language, and time differences.

Culture affects global teams in many ways, from what is acceptable to say to project team members, to how overtime and vacations are used. For example, it is a typically American attitude that if the project is running a week late, project members will forgo or reschedule their vacations. This is not common in European countries.

Communicating can be tricky — especially if not everyone on the team speaks the same language. But even if the languages are the same, what we say may not express exactly what we mean. When everyone is located in the same place, we have many opportunities for informal communications to clarify what we said in person or in e-mail. In global projects, developers have few — if any — face-to-face communication opportunities to clarify what is said.

Communicating across time zones is another challenge. Although working asynchronously can help a team progress faster, not everyone is available at the same time. This may slow down communication and decision making. It's difficult to find common meeting times, whether for project meetings or formal technical reviews. A project manager or team leader working on such a project must balance organizational skills, communication, and tools to make the project work. In my experience leading globally dispersed teams, six rules of thumb have made my teams more productive and effective.

Tip One: Define Complementary Processes and Agree on the Meaning of Important Terms

Global projects are generally composed of teams that do things differently. Some differences are cultural, while others stem from management styles and strategies. What is certain is that each team's reaction to the other teams' processes and terminology won't be the same.

Product development processes don't have to be the same, but they do need to be complementary. By complementary, I mean the outputs of each group's processes should match the expectations of the other groups.

Software development terms vary from team to team as much as processes do. I once led a second-line support group for a globally dispersed team where there was some confusion about what the term "fixed" meant. Our job was to fix the defects that the first-line support group could not fix and were time-critical for our customers. We had a recurring problem with two of our European first-line support groups. They would tell high-profile customers a defect was "fixed" when they took immediate action on it. However, the defects were not necessarily fixed. The Boston group was using the notation "fix" to indicate defects that had been investigated, had a known cause, and were in the progress of being fixed. "Verify" was the notation for completed fixes. It never occurred to our European counterparts that "fix" didn't mean something was truly fixed!

Other terms I've found confusing are "code freeze" and "feature freeze." To me, code freeze is when only show-stopping fixes are allowed. Some teams assume code freeze means only high-priority fixes will get checked in. Others assume most modules will not change, but for some, there is still possibility for change.

It's also important that everyone agree on project milestones — what they are, and what they mean. One of the first times I realized how important it was to understand what the milestones meant was when I was a program manager trying to bring together project components from Boston and Los Angeles. I'd developed the schedule with the participation of the technical leads and their teams. We did fine until the first milestone — feature freeze. To the Boston team, feature freeze meant the low-level design was complete. However, to the Los Angeles team, feature freeze meant they had a pretty good idea of the high-level design. The notion that feature freeze should mean the module interface designs were complete was foreign to the Los Angeles group.

I brought the technical leads together to talk about why we needed freezes, and why and when freezing module interface designs was a good idea. We had all worked together in the same company for at least two years, so we knew each other but had never attempted a cross-country development project. We talked by speakerphone and somehow managed to keep from yelling at one another. It took us about a week to come to a consensus. Not everyone liked it, but it was something we all could live with. We agreed what each milestone meant and the results we would provide to reach those milestones.

I've had several successes with projects using this "global discuss and publish" technique. Some teams chose to define interim milestones in addition to the milestones defined in the overall project plan. By agreeing to what each project milestone meant, we developed a joint project schedule that included the major hand-offs between each group.

When everyone agrees on a definition, it brings all the milestones closer to reality. When milestones are met, they can be jointly celebrated.

Tip Two: Use Configuration Management Systems (CMS) and Defect Tracking Systems (DTS)

When using CMS and DTS, it's important to make sure everyone uses them in the same way. Everyone on the project needs to know where the source files are stored, what their state is, and what can be done with them. Here's a scenario that happened to one of my recent clients.

This organization had development teams in Boston, San Francisco, and Toronto. The project was a collection of fairly independent parts. They all used the same CMS and DTS, but in different ways. The Boston group used a nightly build and multiple promotion scheme for the sources (the first step was to successfully compile and build; the second step was to pass the minimum acceptance test). The San Francisco group used a weekly build and single promotion scheme (it built once a week and fixed things until the build passed the minimum acceptance test). The Toronto group used a scheme based on the review state, not the compile or build state.

This project got into trouble when the integration phase began. Because everyone had different ideas of how to use the CMS, the state of the sources was unclear to the project team. We solved this problem by understanding what each group did with their sources, and how that worked within the overall project; deciding how we were going to use the source tree (complementary processes) so we could actually integrate the whole project; and deciding what the branch and label names meant within each branch and to each group.

You can avoid this problem by agreeing on how to use the CMS and DTS at the start of the project. Using an integrated CMS and DTS tool provides a number of benefits for the project team. For example, the defects and their effect on the sources are clearer to the whole team when they only have one way to look at them. Additionally, the whole workflow of the defect finding and source updating is made easier with an integrated tool.

Not every team wants to use the same tools in the same way, however. I have used the following techniques to get people to work together on this problem:

Logic   I appeal to team members' reason and explain how this will be better for everyone on the team.

Challenge   I challenge the team members to prove me wrong by using the tools for a specified period of time. If they can find a problem during that time, we agree not to use the tools.

Pilot   I request that the team pilot the use of the tools. If members find problems with how tools are used, we reevaluate the tools and how they're used.

Shame and guilt   When all else fails, I fall back on using either a team member's shame of holding the whole team back or his or her guilt about not using the tool to get everyone to at least try what I want them to do.

Using the tools in an agreed-upon way is helpful to developing a complementary development process. Even if the processes are not exactly the same, they should accommodate each other.

Tip Three: Formally Inspect Requirements Documents with All Development Teams

Getting the requirements right is key to project success — no matter what kind of project you're leading. In a global project, it's even more critical. Because it may not be easy to talk to the person who has the necessary information, it's critical to write down, review, and track requirements. It's especially useful to keep requirements in a repository so people can go to one place to continually verify what is going on with the requirements.

Requirements reviews must also be more formal. You can't simply do a casual walk-through with whomever is available or do an informal review over coffee. Formal reviews should include one representative from each team. These participants sign-off that the requirements are correct and ready for their team to implement.

In my experience, electronic-only reviews and inspections are inadequate for effectively reviewing requirements documents. This is true for several reasons. First, the people who initially read the work product direct the discussion. Some are too shy to bring up issues electronically. Second, some people don't read the product once others start commenting. Third, it's difficult to get people to agree on a consistent commenting style. Fourth, some topics of discussion can't be bridged without some verbal contact, and the way some cultures use and understand language to write specifications and comments prevents effective e-mail reviews.

I strongly recommend all technical leaders meet with the marketing leaders and use a Fagan-style inspection in which everyone reads each line of the requirements specification together. Video and web technology can make these reviews and inspections easier. A number of Internet and web-based products let you share HTML and non-HTML documents among a number of sites.

Electronic whiteboards can be particularly useful if you need to discuss design or architecture issues and draw pictures. Normal video communications may be most useful for standard project meetings, rather than meetings focused on carefully reviewing a technical document for defects.

You can use the same Fagan-style inspection process on architecture specifications, design specifications, and test designs (the strategy and plans for testing the product). The project can gain something valuable from inspecting test designs with representatives from each team. This way, each team's perspective on the product and what is important to test is discussed. (Discourse continues on page 34.)

Integrated CMS and DTS Tools

You can avoid communication problems and other pitfalls of globally distributed development with configuration management, requirements management, and conferencing tools. Here are a few to check out.

Configuration Management (CM) and Project Management (PM) Tools
ClearCase (CM) and ClearTrack (PM)
Rational Software Corp.
20 Maguire Rd.
Lexington, Mass. 02173-3104
Tel: (617) 676-2400
info@rational.com

MKS Source Integrity (CM)
Track Integrity (PM)
MKS Inc.
185 Columbia St. W.
Waterloo, Ont.
Canada N2L 5Z5
Tel: (800) 265-2797
inquiry@mks.com

PVCS (CM) and PVCS Tracker (PM)
Intersolv Inc.
1700 N.W. 167th Pl.
Beaverton, Ore. 97006
Tel: (503) 645-1150
pvcsinfo@intersolv.com

Continuus/CM
Continuus Software
108 Pacifica
Irvine, Calif. 92718
Tel: (949) 453-2200
www.continuus.com

For more vendors and configuration management tools, visit www.iac.honey-well.com/Pub/Tech/CM/CMTools. For more project management tools, visit www.iac.honey-well. com/Pub/Tech /CM/PMTools.html, or check out the project management buyer's guide at www.sdmagazine.com.

Requirements Tracking Tools
Requisite Pro
Rational Software Corp.
4900 Pearl E. Cir., Suite 106
Boulder, CO 80301
Tel: (800) 732-4047
info@rational.com

Doors
QSS Inc.
200 Valley Rd., Suite 306
Mt. Arlington, N.J. 07856
Tel: (973) 770-6400
info@qssinc.com

For more information on requirements tracking tools, you can visit www.incose.org/workgrps/tools /contacts.html.

Internet and Web-Based Products that Let You Share HTML and Non-HTML Documents
WebLine Communications' WebLine lets you "push" web pages to other people's machines. For more information, visit www.webline.com.

Microsoft's NetMeeting lets participants share documents during a meeting. For more information, visit www.microsoft.com/netmeeting.

Swift Computers' SoftBoard and TeamBoard's TeamBoard are electronic whiteboards that let people create and modify documents electronically. For more information, visit swift@neosoft.com and www.teamboard.com.

Continuus/WebSynergy supports the collaborative development, management, approval, and deployment of all types of web content and software. For more information, visit www.continuus.com.

Tip Four: Provide All Team Members with Project Plans

Project leaders sometimes forget that not everyone has access to or knows all the intricate pieces of the project schedule. In a global project, this can lead to project failure. Once the project plan is developed, everyone needs access to it. Joint development of the project schedule will ensure all the hand-offs and milestones are well understood and articulated by everyone. At the very least, I recommend the major milestones and their commitment dates be pulled out of the schedule and disseminated to the entire global team by e-mail. It is even better to have the whole schedule and project plan available online in a workgroup tool.

In addition, I recommend that every two weeks, or more often in "crunch mode," you send out an updated list of major schedule milestones to everyone in the company or everyone on the project. This helps everyone see which milestones and dates have been reached and tracks the team's progress on the project. As each group meets its milestones, send out congratulatory e-mail to the specific project team, copied to the other teams. This gives the other teams a chance to stay informed about how the project is doing.

Tip Five: Organize Project Teams by Product Feature

I've seen global teams organized by product development function and by product feature. Although it's possible to have developers in one place, writers in another, and testers in a third, they may find it harder to do the actual work of product development. I worked with one functionally organized group where the developers and testers were in Massachusetts and the support group was in California. The two groups had dramatically different perceptions of what each other did.

The support group sent detailed bug reports to the developers for fixing in the next release. In addition, some defects needed to be fixed as updates to the current release. But not all the reported defects were fixed in a timely fashion. The developers didn't always understand the priority of the fix requests, and the support group didn't always explain the importance of the customer. As a result, the developer and support groups weren't able to work smoothly with each other.

In my experience, global projects are more successful if teams are organized by product feature. One of my clients currently working on a component project has organized the components into three related areas. Two U.S.-based teams are working on two areas, and one European team is working on the third area. These teams are self-contained — they each have product developers, product writers, and product testers. The system test group is part of one of the U.S. teams, and that group will have final responsibility for complete system test. At this point, the people understand what they have to do, and the project is progressing on schedule.

Tip Six: Use Collaborative Tools to Bring the Project Together

Especially in a global project, collaborative workgroup and workflow tools let people see all of the documents in one place. Workgroup and workflow tools such as Lotus Notes help bridge the communications gap of time and language and lessen the effects of cultural differences on processes.

For example, during the project with the Boston, San Francisco, and Toronto teams, I used forms in Lotus Notes to keep senior management and members of other projects updated on our project progress and to get help from a centralized support group with our

CMS. At one point, senior management wanted to add another project to the product. Because the project managers and senior management had access to all the project plans and release criteria, we could rationally discuss how to add the new project and still meet our ship date.

On this same project, we used workflow forms to create and inspect the master version of the product. We sent the release group a workflow form explaining how to create the master. They filled out the next form telling us where the master was and what had gone into it so we could verify it. We replied to that form with the results of our testing. Using the workflow and workgroup parts of the tool saved us significant time on the project by letting us work asynchronously in time, but together in process.

Global projects pose additional challenges to the normal stresses of software development projects. Different cultures, languages, and time zones all contribute to the extra stresses. By keeping these six tips in mind, and by applying all the sound project management practices you've already learned, you can keep your global projects on track.

2.5.2 "Managing Collaborations"

by Mary Loomis, January 1998
Edited by Larry Constantine

Software projects that span multiple organizations are becoming more common. However, they also present their own people management challenges.

After a recent conference, one of the industry leaders in object-oriented software shared with me his fantasy of a future software development think tank. This enterprise would bring together some of the field's top innovators in practices and processes to extend the leading edge in software development. What did I think? It sounded great, but...

A think tank requires thinkers: creative, energetic, and independent leaders who can push the envelope. The problem is that such innovative thinkers do not readily collaborate. Just picture what would happen if you brought together some of the top names, the top minds in the industry and put them in the same office. Managing such a collaboration could be Sisyphean. Who could lead a bunch of leaders? My colleague immediately thought of somebody — a manager for whom he himself would happily work, one with the skill and finesse to handle the position: Mary Loomis.

After such a high recommendation, it seemed appropriate to invite Mary to contribute some of her insights into managing collaborations. — Larry Constantine

Managing the isolated software project team can be quite a challenge; many of us bear battle scars we've earned in the process. Managing collaborations across multiple organizations involves further complications. Such collaborative efforts are becoming increasingly common. Quite often, my success as a manager is largely determined by the ability of the people in my organization to collaborate with people in other organizations. Here are some of the guidelines I've gathered and developed to increase the probability that such a collaborative software project will be successful.

In a collaboration among multiple organizations, diverse parties work together for some period to achieve a common goal. Diversity comes from having multiple organizations with multiple management chains involved, sometimes in different companies. Because each organization introduces its own motives, culture, and reward system (not to mention personalities and egos) into the software project soup, multi-organizational projects are especially challenging to manage.

Real energy is needed to sustain cooperation when organizations work together, but a team member should not plan to make a career out of his or her participation in the collaboration. The joint effort has a finite life span, and the people involved often need to continue to pursue their separate efforts in parallel with working on the collaboration. The life spans of collaborations are relatively short. Even when the lifetime is measured in years rather than weeks, collaborations are temporary assignments.

The goal of collaboration is to produce some software result, ranging from a specification to a running system. There must be valid reasons for working together toward this goal: some identifiable rationale for forming the collaboration and its management, and there must be participants willing to deal with its inherent complexities.

Here are some characteristics I've observed during successful collaborations I've been involved in. They're listed in no particular sequence; all are important:

Leadership

A successful collaboration has a leader who is respected by the entire team. Sometimes two co-leaders can work together, but they must trust each other and communicate well. The collaboration will not work without an up-front agreement as to how to resolve disagreements. For example, in one collaboration between my organization and another, the other manager and I alternated the lead role over both organizations monthly. Each of us trusted the other to make decisions on behalf of the overall collaboration.

As a counterexample, another collaboration in which my team was involved had no clear leader. Sometimes the marketing group was in the lead (and making commitments to pilot customers), and sometimes the development group was in the lead (also making commitments to pilot customers). The result was over-commitment, dissatisfied pilot customers, and unhappy team members.

Goal

A successful collaboration has a clear goal and well-specified expectations. All the team members understand why they are working together. Team members frequently assess their progress toward the goal together. Using the team members' commitment to pursue the goal, the leader must make sure the goal is highly visible to keep the effort on track. If the goal is obscured, collaborative efforts will easily get bogged down in extra work.

Without a clear goal, it is difficult to answer questions like, "What impact will not doing x have?" It can be difficult to detect when group discussions are headed down the proverbial rat hole. It's also difficult to detect when groupthink is leading a team on an undesirable path. I like to reinforce context by starting every planning or status meeting with a brief reiteration of the collaboration's goal.

Roles

In successful collaborations, each team member has a well-defined role and trusts others to fulfill their roles. Without this trust, team members may encroach on each other's territory in attempts to get the overall task done.

I once painfully survived a collaboration in which roles were stated but weren't very clear. One application team needed a fix to the infrastructure code, but didn't trust the infrastructure team to be responsive. The application team decided to make the fix itself, but didn't communicate this with the infrastructure team. During the following months, the application team frequently complained that it was overworked, relative to the infrastructure team, which apparently had less work to do. Territorial encroachment can threaten a collaboration's chances of success.

Buy-in

A successful collaboration has buy-in from all appropriate levels in all the participating organizations. One technique that works well to ensure continued buy-in is to hold reviews of intermediate results, inviting the peers in the involved organizations — especially within the same company — to attend the same review. For example, an architectural review of an effort intended to provide an integration path for multiple software product lines might include the division managers from all the affected divisions, rather than conducting a separate review for each. This approach increases the opportunity for everyone to understand trade-offs and different organizations' perspectives.

It is also important to identify not only the stakeholders who have resources invested in the effort, but also the managers who can make strategic decisions that could torpedo the effort. The team should identify these people and communicate with them, asking them to publicize their endorsement of the work, both within their organizations and to the team. I know of a collaboration that involved several organizations and had many levels of managers involved. Nearly all peers communicated openly and well, except at the second-level manager level. Mixed signals were getting to the team members, and priorities became confused. Then the torpedo arrived: a high-level manager of one of the organizations dramatically changed that division's business strategy, making the success of the collaboration essentially irrelevant to the division. I don't know what to do to avoid this kind of situation, other than be vigilant for early warning signals and keep testing for continued buy-in.

Schedule and Interdependencies

A successful collaboration has a clearly visible schedule, with reasonable delivery points and explicit recognition of dependencies across organizations. There is a process for handling changes and for tracking status. A collaboration in trouble has no schedule reflecting the combined efforts of all the participants. Watch out also for schedules that bear little resemblance to reality. Ensure that first-level managers and engineers agree that schedules make sense. I like to reinforce the notion that schedules are living documents. Tracking progress relative to a schedule helps the team know when either the schedule needs to change or something about the team's operation has to change.

Resolution Process

A successful collaboration has a well-defined process for surfacing and resolving issues. It is important for the collaboration's leader to design the process before the issues begin to arise. The team needs to understand the process, and it must fit the cultures of the organizations. Organizations with open, nonthreatening environments may be able to expect individuals to step forward to highlight issues, either spontaneously or in periodic status meetings. Leaders in more closed organizations need to actively seek out issues without putting individuals on the spot in group settings.

Communication

A successful multi-organization collaboration has a well-designed system of communication and communication tools. Typical tools include regular meetings, shared workgroup databases, e-mail and voicemail distribution lists, shared sets of presentation materials, source code control systems, and document version management systems. The tools matter less than the way the team uses them.

Meetings may take place in person or via teleconferencing or video conferencing, with notes distributed to team members after the meeting. A common set of tools can make a huge difference to the operation of a collaboration.

Regular communications and flexibility are especially vital in collaborations among geographically dispersed teams. When teams are many time zones apart, finding mutually acceptable meeting times can be a challenge. The team should either find an acceptable time and use it consistently, or move the time around so everyone is equally inconvenienced. Meetings should be efficiently run, and it must be clear that participation is mandatory.

Vocabulary and Mind-set

In a successful collaboration, team members understand what other team members are talking about. It is worthwhile to invest time to develop a common terminology and vocabulary. Requirements and design models are important to nearly all software projects, but especially for collaborative software development. These models help clarify the semantics of the project area and can be instrumental in helping team members understand each other.

However, vocabulary is simply not enough. In a successful collaboration, the parties understand each others' frames of reference, personal motivations, and business reasons for being involved. To make sure these are clear, I like to ask my team members to explain the viewpoints of people from other organizations.

Credit

In a successful collaboration, there is plenty of credit given for everyone's efforts. Team members should openly acknowledge each other's contributions, especially across organization boundaries. Managers should give credit to the other organizations. After all, together the collaborators are expected to perform better than any of the individual organizations could alone.

Extensive credit-giving works best when practiced by all the participating organizations. Even if you find your organization on the short end of the credit stick, I suggest you not scale back. As a manager, your role is to support the entire effort and ensure overall success, not to grab charter or to compete with your collaborators. In my experience, people notice managers who don't give credit where credit is due; it doesn't help morale at all.

Don't force it. Don't collaborate merely for the sake of collaborating. Forced situations are typically doomed. Such collaborations introduce more complexity than value. The effort might better be pursued independently by one — maybe any one — of the organizations.

Software Collaborations Are Different

These characteristics of successful collaborations are mostly people-related and are probably not unique to software collaborations. What makes software collaborations different? A few factors are worth noting.

First, it can be difficult to measure the progress of any software project, especially a collaborative one. Managing interorganization dependencies and risks is complicated substantially by the uncertainties seemingly inherent in accurately predicting the schedules of software projects. I have found that the project schedule is the key tool in managing these uncertainties.

Unlike many other industries, the software business does not have large stores of tested, standardized parts to draw from in constructing new systems. With-out standardization, communication of the details becomes even more essential.

Another difference inherent in software collaborations seems to be the extent of interdependencies. Unmanaged interdependencies have a way of showing up later as bugs, when one person's efforts have side effects on the work of others. Management attention and careful partitioning, both of the software and of people's roles and responsibilities, can help enormously.

The software business also lacks the kinds of blueprint languages typically used to ensure valid communication in other kinds of collaborations, for example, in designing and implementing buildings, computing hardware, airplanes, automobiles, or toasters. No matter how conscientious we are about specifying software interfaces, functionality, or expected quality of service (performance, reliability, security, and so on), we're hardly ever precise enough.

Unlike many other industries, the software business does not have large stores of tested, standardized parts to draw from in constructing new systems. Without standardization, communication of the details becomes even more essential. A modeling language, like the Unified Modeling Language adopted by the Object Management Group, is the closest thing we have to a software blueprint language. Managers can lower software project risks by requiring sufficient investment in modeling. Models document the architecture for the project.

Like many principles of management, the characteristics of successful collaborations may seem pretty obvious when captured in one place. Yet we often have trouble practicing them in the real world. Perhaps there are other guidelines you'd like to contribute to the list. I'd really welcome hearing your ideas. Maybe we can collaborate.

2.5.3 "Implementing Feature Teams"

by Jim McCarthy, December 1995

In this excerpt from his book, Dynamics of *Software Development*, Jim McCarthy discusses the multidisciplinary teams that have served Microsoft's Visual C++ group so well.

In the Visual C++ group at Microsoft, we use feature teams. I am unabashedly enthused about feature teams. They've had an amazing effect on our group. If you ask a quality assurance person from our team, "What's your job?" he or she will tell you, "My job is to ship the product." Not to test the product — that's a myopic view. The quality assurance person's job is to ship the product. The quality assurance person's job is to develop the product. The quality assurance person's job is to design the product, to know the customers, and, in fact, to know everything that's going on in our technology plan, our product, our market, and our business.

We organize the development team — Quality Assurance, Development, Documentation, Program Management, and, as appropriate, Marketing — in a conventional hierarchy, albeit a rather flat one (at least in the best cases); but we also take representatives from each of those four or five groups and put them into a feature team in a matrix organization. We take a chunk of the product, say, foobars, and tell the feature team, "OK, you are the foobar team." And they define the foobar as a team. They create their milestones as a team and articulate their schedule. They even define their process as a team — although we're always encouraging common processes across teams.

Feature teams are about empowerment, accountability, identity, consensus, and balance. They are the best mechanism I have encountered for remediating (over time) organizational "goofiness."

Empowerment

While it would be difficult to entrust one functional group or a single functional hierarchy, such as Development, for instance, with virtually absolute control over a particular technology area, it's a good idea to do that with a balanced, multidisciplinary team. The frontline experts are the people who know more than anyone else about their area, and it seems dumb not to find a way to let them have control over their area. Although I can think of many bad decisions (and good ones) made by managers and overturned by other managers, I can't recall a single case in which the feature team was overruled by management muckety-mucks. I doubt whether at this stage of our group's evolution such a thing would even be possible. This says something about both the authentic power feature teams can wield and the quality of their output. It also says a great deal about the importance of good hiring practices.

Accountability

One of the most interesting software development-related discussions I ever had was with the leadership of a feature team on the topic of accountability. The question we discussed was "What are you accountable for?" My theory is that most of the best thinking in any group of humans is wasted, lost, because people who don't feel accountable can get into neurotic ways of responding to new ideas.

First, the source — the person or group having the thought or the idea — might feel "It isn't my area of responsibility" and so never say anything at all in a constructive way. The subsequent variety of pathological acting out of the critical thought can be astonishing. The "thought never spoken" gets enacted somehow, leading to all kinds of confusion, suboptimal behavior, and expense. Teams will vote with their feet if they don't talk with their mouths.

Second, if the source does say something (which is tremendously risky), the person or group at whom the thought or idea is directed (the target) can react defensively, and there are infinite styles of defensiveness. Sometimes, defensiveness is quite hard to recognize, particularly in groups in which defensiveness has been identified as a negative value. Ironically, a more evolved group, one that has an ongoing consciousness of defensive behavior, simply evolves more elaborate and subtle ways to defend. Vociferous support of the status quo by the person or team responsible for it should always attract analysis.

Third, the source of the original idea might not stick with it until the idea detaches from the ego and becomes available to the target. Just as defensiveness is prevalent, so, too, is aggression. Usually, the initial attempts to voice concern are all wrapped up in the built-in aggression and judgmentality that is nearly universal among humans. That triggers (appropriately) the defensiveness mentioned above. The target correctly perceives the aggression and, in defending against it, misses the "noble" component of the message. At this point, the source succumbs to its own aggressive pathology and dismisses the target as unworthy of further consideration and risk — that is, BOZO = TRUE.

Thus are many good critical constructs lost. The only remedy I know for this syndrome is the workshop. In a workshop environment, people voluntarily step forward and solicit the critical perceptions of their colleagues. In that single gesture of asking for feedback, they (more or less) disarm. This receptivity goes a long way in nullifying to a large extent the aggressive pathology of the source. And, because each person or group is a potential target of all the others, the ritual disarming pays everyone. I'm surprised that we don't use the workshop technique more frequently and more broadly.

I have successfully applied the workshop idea to real-world software development problems by promulgating the notion of "case studies" among some of the teams in my world. It works like this. A person volunteers to deliver an account of a case. The case, of course, is a current (or at least recent) professional and usually interpersonal situation in which the person is embroiled — for example, "I can't get buy-off for this idea," or "Nobody reads my mail," or "Developer X won't stick to the plan." The person talks for a few minutes about the case, sometimes changing names, sometimes not. The rest of the team then questions the volunteer for more information and clarification — what steps have been tried, for instance. Then a somewhat prolonged discussion of the case ensues.

Two things about this simple technique never cease to amaze me. First, a plethora of ideas is generated. You can barely keep count of the times the volunteer says something to the effect of, "That's a good idea" while scribbling furiously in his or her notebook. Second, the case almost always gets generalized. The rest of the team will say things like, "That's just like problem x I had last month." After enough similar cases have been mentioned, someone will say, "Isn't your problem really just a special case of y?" They will have identified the general principle or the general situation. Then all scribble in their notebooks.

Accountability offers lots of advantages and very few limits. If a balanced group of people is mutually accountable for all aspects of design, development, debugging, quality assurance, shipping, and so on, they will devise ways to share critical observations with one another. Because they are accountable, if they perceive it, they own it. They must pass the perception to the rest of the team.

Defensiveness is hard to recognize, particularly in groups in which defensiveness has been identified as a negative value. A more evolved group simply evolves more elaborate and subtle ways to defend.

Identity

Empowerment and accountability lead inevitably to identity. People identify with what they can control or influence. The higher the degree of control, the more the identification. While this kind of identification is generally good and is at least a prerequisite for great software, it can lead to situations in which gross individual pathologies express themselves in the product. An individual's self-destructiveness, for instance, might show up in the way he or she handles the function he or she identifies with.

With cross-functional feature teams, individuals gradually begin to identify with a part of the product rather than with a narrow specialized skill. There are very few competitors left to blame, and one's own naked success or failure becomes painfully clear. You can't go on blaming a management specter when all your encounters with management are of the kind in which management says, "What more do you need to achieve your goals?" You can't blame other functional areas when they are mutually accountable with you for all aspects of your feature. The question becomes "What's blocking you?" In an empowered environment, the answer is inevitably something within.

Consensus

Consensus is the atmosphere of a feature team. Since the point of identification is the feature rather than the function, and since the accountability for the feature is mutual, a certain degree of openness is safe, even necessary. I have observed teams reorganizing themselves, creating a new vision, reallocating resources, changing schedules, all without sticky conflict. The conflicts that do arise tend not to be personal and are generally resolved without recourse to management.

I don't recall a single case in which the feature team was overruled by management muckety-mucks.

Balance

Balance on a feature team is about diverse skill sets, diverse assignments, and diverse points of view. Because the team is composed of people representing multiple functional disciplines, it is diverse prima facie with respect to skills. Usually, people who have specialized in one or another skill also have points of view that are complementary to their discipline. We can understand the diversity of team assignments by looking at the team's diverse points of view, characterized by the kinds of questions each team member type might be expected to worry over.

Program Management   What is the state of the team? How is the process going? How effective is our leadership? Where are we in the cycle? What is the quality of the schedule? Have the things we needed from other teams come in? Are our goals clear? What are we kidding ourselves about today? What list of things needs to be accomplished now? What is the theme of team behavior this week?

Quality Assurance   What is the probability that I will get the next hand-off from Development? Where are we going with bugs? What types of bugs are we seeing? What functions are simply not there yet? How does this piece perform? Should I be raising any red flags this week? Does Program Management know how stable (or unstable) this piece is? How are our communications? Does everybody share a common point of view with respect to our state? What is the reasonableness of our stated goals for this week?

Development   What is the probability that I will meet this week's deliverables to Quality Assurance and Documentation? Is this piece well designed? Will the user get it? Is how to use it clear? Is it fast enough, small enough? Have I fixed all the bugs in this piece? Is anything I'm doing or not doing blocking anybody else on the team? Do I believe in this week's goals? Is this work on or off strategy? Does it contribute to the technology plan or is it a throwaway?

Product Management and Marketing   What will the customers think about this feature? How can I make the feature vivid to them in my communications? What emotional associations does this feature trigger? How could we intensify those associations by changing the product? How will the customers feel when they hear about this feature from me? When they first use it? Does this feature as presently constituted advance our relationship with the customers? What will the press make of it? What is the story behind this feature? Do I totally understand what the feature is and what its significance is? What is the probability of schedule success? Should I be introducing more uncertainty into the rest of the organization? Or more certainty? Where are we on the certainty-o-meter?

Documentation and User Education   Can this feature work any more simply? Is my explanation of how to use this feature utterly clear? Can we make it so that no explanation is required? So that the feature reveals itself to the maximum? If I can't write about the feature yet, is it late? Does Quality Assurance think the feature is done? How does this feature feel to me? Is there any possible way to better it? Can I express myself more succinctly? Does the rest of the team "buy in" to my writing?

Even though they're immensely valuable, feature teams can be hard to get together. Moving from a conventional hierarchical structure to the successful application of feature teams can be a very difficult transition. Several kinds of difficulties can arise from within and outside the feature team.

Inside the feature team, there will be a great deal of uncertainty. The feature team members won't know the limits of their autonomy. They'll feel that they ought to be working together better in some undefined way since management clearly wants them to be a "team," whatever that might mean. People will cling to their traditional roles. Developers will naturally be the dominant players at first. It will seem as if everybody on the team is playing his or her original role with only marginally improved team communication. In fact, the cost of assembling the team, having a few meetings, deciding who is on which team, reorganizing the group, and so on will at first seem out of proportion to the benefit of marginally improved communication.

People don't "get" feature teams at first. It's only as the inevitable failures start to mount up that the opportunities for bonding become manifest. Often a team will feel frustrated and abandoned as Management leaves the feature team alone to decide its own fate. This autonomy will produce considerable discomfort because people are unused to providing their own authority.

There will also be a great deal of conflict aversion, and this will be a continuing problem because most people's superficial notion of "teamwork" is that it is equivalent to some namby-pamby consensus and bogus good cheer.

The only consensus worth having is a creative one achieved in the combat of fully engaged intellects. Such a consensus is born of sleepless nights, fear of rejection, and trials of personal courage. Conflict, which usually presages growth, is the hallmark of such consensus.

But initially, the feature teams are lackluster. The question of courage never arises because the challenge is to simply endure this latest management fad. Eventually, it will begin to dawn on some of the team members that they are, in fact, in power. No one is holding them back.

Resources are available. Creativity is welcome. Management is there only to support their goals. They're within striking distance of the goals that started them on the course of software development in the first place. In a penetrating and even earthshaking moment, they realize that they are free to do and accountable for doing.

This creates another set of problems, of course. Many creative and brilliant people unconsciously insist that something hold them back, that some negative force prevent their gifts from emerging. They carry around within them a "governor" function that blocks the ultimate release of their full creative energy. This self-inhibiting stems no doubt from some early parental rejection of the child's beauty, passed blindly from parent to child, some introjected fearfulness of being: if I truly reveal my uniqueness, the child senses, you (the parent) will abandon me.

Since few parents ever explicitly demand that their child limit his or her growth, we all tend to develop extremely subtle sensitivities that enable us to detect these negative parental demands. There probably is some healthy, gene-protecting impulse behind this urge to be "normal," to be the same as everybody else, to gain broad acceptance in the mediocre, average community, to conform to some homogenized value system; but this impulse is the antithesis of what is required for intellectual leadership and for the creation of great software.

Our self-limiting sensitivities — our compulsive search for negation — although healthy in a pathological environment become pathological in a healthy environment. We've attuned ourselves to perceive signals at every turn from our "management" and other authority figures that we are not to engage to the absolute fullest. And Management might in fact be unconsciously putting out these "stop" signals (in the same blind and tragic way that parents undo their own children), but in the main, Management at this stage should be (and usually is) a benign and affirmative supplier. At the very least, a good faith effort is usually in place.

People's superficial notion of "teamwork" is that it is equivalent to some namby-pamby consensus and bogus good cheer. The only consensus worth having is a creative one achieved in the combat of fully engaged intellects.

As individuals begin to "get" (and it is an "ah-ha"-grade experience) the fact that no one is opposing them and that they are accountable for their own success, they invariably transmit this notion, this feeling of liberation and empowerment, to others on the team. This personal liberation will be understood in varying degrees within a broad range of moments and manifested in a variety of ways, individual by individual. It's fascinating to observe. Each person needs to be challenged, encouraged, and nurtured. Flexibility and patience are demanded of everyone as people grow into their own awareness at their own paces. This growth process, however difficult, is a prerequisite for greatness in any intellectual team endeavor.

I've also observed that the degree to which people from a functional discipline (say, Documentation) will fully participate in the feature team is inversely proportional to their degree of success in the old, less-empowered regime. If a relatively functional team had isolated itself from the widespread dysfunctionality that had been surrounding it in the previous regime, and if it had been more or less successful as a result, the members of that older team will have a difficult time undoing their isolationist maladaptation when the rest of the organization grows healthier and it becomes safe to "come out and play."

Outside the feature team, functional area managers (development managers, quality assurance managers, manager managers, and so on) are meanwhile engaged in their own, often private, struggles. In spite of the lack of creative and intellectual teamwork in the overall group, they have succeeded as individual managers before "this feature team craze." To at least some extent, they will necessarily be opposed to the growth of the team. After all, they rose to positions of responsibility and authority in the unhealthy environment. They have been selected from the pool of available talent because they could achieve some level of results in spite of the handicaps placed on them by the broader dysfunctionality. Now they are faced with the increasing irrelevance of their core skills. Their ability to empower themselves in a generally disempowered environment is no longer needed in an environment in which broad empowerment is validated. What has gained them success to date is now seen as a maladaptation. That which they were heretofore rewarded for becomes something they are penalized for.

These managers know how to ship software in a world opposed, how to get things done in an intellectual and creative ghetto. They watch the first tentative steps of the feature teams with concern, even alarm: "But they're doing it wrong. I should be doing this. I should just tell them what to do." These reactions come out of genuine concern. The temptation to "take charge" is nearly overwhelming. The senior manager will fear at first (I certainly did!) that the whole feature team thing may be crazy, that letting people with less experience and judgment decide the fate of the technology and the business is absurd, that they're making mistakes the manager has already made!

As the various managers learn to walk the line between encouraging and teaching on the one hand and commanding and controlling on the other, bizarre conversations will ensue about who should be deciding what, about what the managers' roles are if not to control.

I experienced doubt. Several key managers were wondering what in God's name we were doing to a perfectly good team. I would wake up in the middle of the night, wondering why on earth we'd set out on this crazy jihad to get everybody's head into the game. We managers ultimately referred to this period as the time of our "Feature Team Angst." Gradually, we began to see that our role was to teach, to challenge, to encourage, and to add whatever value we could to the process. If our ideas were the best, they would survive. While the power resided in the main with the feature team, we could challenge their assumptions, get them to examine their motives and behavior, help them reach consensus and manage conflict, and articulate what we understood their vision to be back to them. We could coach them on their effectiveness. It was essential, however, because we wanted accountability from the teams, that we endow them with the power to decide and to act as they saw fit.

As various managers learn to walk the line between encouraging and teaching on the one hand and commanding and controlling on the other, bizarre conversations ensue about who should be deciding what, about what the managers' roles are if not to control.

Of course, this entire process was gradual, and it is to some extent continuing to develop. There were many occasions when the feature teams basically demanded that management make some decisions or set some goals; likewise, there were some times when management inappropriately mandated something or other to the teams. It has not been a pure situation.

But in the midst of all the change, I can occasionally catch glimpses of the ultimate organizational strategy for creating intellectual property.

In the ideal project, there are basically Creators and Facilitators. The Creators specialize in development, marketing, quality assurance, or documentation or some other directly applied software discipline. The Facilitators specialize in group cognition, in creating an environment in which it's safe to be creative and in which all the resources needed are applied to the problems associated with making ideas effective. The Facilitators make sure that the maximum number of ideas shows up "in the box." The Facilitators evolve from the ranks of today's managers and program managers, the people whose skills show up in the box indirectly.

The Facilitators are just as accountable for the quality of what ends up in the box as the Creators; and the Creators are as accountable for the extent of group cognition as the Facilitators. It's a good idea for these groups to be held accountable to each other, to be "reviewed" by each other in some fashion. The old hierarchy diminishes in importance. The only authority stems from knowledge, not position. This is a very challenging vision.

2.5.4 "Tips for Technical Interviewing"

by Barbara Hanscome, September 1996

What some of the top software managers ask and look for when hiring technical staff members.

By the time software developers interviewing for positions on Microsoft's Visual C++ team met with the team director, Jim McCarthy, they only thought they had been through the ringer. Sure, they had passed muster with several Microsoft developers and proved their coding skills in lengthy programming evaluations. But McCarthy wasn't interested in their technical know-how. In fact, he asked them only one question. "Tell me why I should hire you. You've got ten minutes to sell me."

Then, McCarthy would sit back and proceed to criticize every word the candidate uttered, barking that their answers — whatever they might have been — were unconvincing, canned, even insincere. "I'd really try to rattle their cage," says McCarthy, who has since left Microsoft and written the book *Dynamics of Software Development* (Microsoft Press, 1995). "I wanted to make the situation as complex as possible and see how the candidate fared in it." More important, McCarthy wanted to break the person out of the "interviewee" role and show some vulnerability. "Once you've got the real person out and you stress them, you'll see how they'll behave during the stress of creating products; like if they'll have the good sense to ask for help, or if they'll be able to admit that they don't know it all."

Every manager has a different approach to interviewing, but most will agree that interviewing software developers is tricky. The skill sets required for good team members — be they programmers, analysts, designers, team leaders, or testers — are large and complex. You're not looking for one skill, but many, including the ability to work well within a team. Some of the things you must assess in an interview include:

• How well candidates know a particular technology or development skill

• Their ability to use those technologies and skills

- How fast they learn new technologies and develop new skills
- How well they work with other people.

Rapid changes in technology make interviewing for specific skills even more difficult. "It's very hard to keep abreast of the changes in technologies and tools, and to hire people who have kept abreast of those changes," says Mary Levine, a senior information technology manager with Charles Schwab in San Francisco, Calif.

Steve Maguire's Interview Tips

For the coding part of the interview, I ask programmers to write C's standard `tolower()` library function. On the surface, this problem might seem too simple to be an effective test of a programmer's skills, but even simple problems are full of implementation possibilities that can be explored.

Once we have code in hand, I start asking questions. Does the code work only for upper-case letters? Does the code unnecessarily return an error if the letter is not upper-case? Does the range check have off-by-one errors? What values would the candidate use to test the code? Does he or she test the boundary conditions, or use ineffective tests such as capital A, B, and C?

How would the programmer make the code faster? How would he or she know the new code is faster? Does the programmer suggest comparing the new code with a timer, or just assume the code is faster because "it's obvious?"

When pressed on the speed issue, does the candidate come up with a macro (or inline) solution, eliminating function call overhead? Does the person know the steps involved in function call overhead? Can he or she describe the typical size-speed trade-offs surrounding macro usage?

What if `tolower` was getting passed lower-case letters 98% of the time? Does he or she use short-circuiting or back-to-back if statements to implement the range check? What if there were gobs of memory? Does a table lookup solution come to mind? Can the candidate explain the pros and cons of using table lookup solutions?

Throughout the interview, I ask as many questions as possible to get a feel for the programmer's breadth of knowledge, and—more important—for how the person thinks. With such an interview, I can get a good feel for a programmer's skill level, including fundamental design skills. It's certainly much better than asking nothing at all.

—Steve Maguire, author of *Writing Solid Code* (Microsoft Press, 1993), and *Debugging the Development Process* (Microsoft Press, 1994).

Defining Your Interviewing Process

The first step of any interview is defining the qualities you're seeking. "The finer grained your search criteria is, the more efficient you can be throughout the process," explains Gene Wang, CEO of Computer Motion in Santa Barbara, Calif., and author of *The Programmers Job Handbook* (Osborne McGraw-Hill, 1996). Some areas to consider include:

- Programming language expertise
- Operating system expertise
- Application expertise

- Team communication expertise
- Size-of-project expertise
- Project ship track record.

At Wang's company, interviews are coordinated by a hiring manager who is usually someone on the development team. This person screens the résumés against the criteria defined for the job. After candidates are screened for matching qualifications — first by résumé, then by phone — the hiring manger sets up personal interviews.

The second interview takes a full day for the candidates, each of whom may end up talking with up to six people. "In the second interview, we start selling the company," says Wang. "Clearly, with the best people, we are just one of many opportunities. So, we need to start putting our best foot forward." This job usually goes to the hiring manager, while the other interviewers continue to evaluate the candidates.

"It's a good idea to have an agenda," says Wang, to hand to the candidates when they arrive for the interview. The agenda lists the names of everyone the candidates will meet and the time frames set aside for each interview. "This level of organization helps the candidates navigate through the day and remember who they're talking to, and it doesn't make you appear like a bumbling idiot."

Andersen Consulting, a management and technology consulting firm based in Chicago, Ill., will hire approximately 3,500 technical professionals this year worldwide, according to David Reed, director of recruiting for the Americas. Andersen's technical interview for programmers and programmer analysts focuses on a combination of system building fundamentals, development life cycle or methodology, and application operations and support.

The programmers and programmer analysts who pass the technical interview, reference checks, and interviews with other Andersen employees next undergo a behavioral assessment where they are measured against a set of characteristics based on traits gleaned from Andersen's top performers. The criteria differ for all positions, but include things like creativity, ability to take initiative, interpersonal communication skills, and intellect.

Probing for Technical Knowledge and Skills

One of the biggest challenges for managers when conducting a technical interview is assessing someone's level of skill in a particular technology. How do you know if the person you're interviewing understands, say, object-oriented programming or Java as well as he or she claims?

Some companies require programming candidates to audition for them. "A software project manager should observe first-hand how a candidate software engineer plays the game before a commitment is made," writes Roger Pressman in his book *A Manager's Guide to Software Engineering* (McGraw-Hill, 1993). "Prior to the day of the interview, the candidate should be asked to bring a portfolio of her or his best work. The portfolio should include descriptions of the application, samples of actual deliverables, and a summary of the end results. The candidate should also be asked to prepare a 15-minute formal presentation on a technical topic related to current work."

Steve Maguire, author of *Writing Solid Code* (Microsoft Press, 1993) and *Debugging the Development Process* (Microsoft Press, 1994) used a slightly different approach when he worked at Microsoft. Maguire gave each candidate a coding problem, then sat down and solved it with them. The problems were simple, such as:

- Can you write a function that counts the number of bits set in an integer?
- How would you traverse a binary tree?
- Given a 100-element sorted array, write a function that returns the index where the integer "I" is first found.

Each problem had many possible solutions, however. "The question must be extremely simple and straightforward," explains Maguire. "It's very important that it not be a trick question, because then the candidates get so confused they can't possibly demonstrate their skills." Maguire watched carefully how the candidates attacked the problem. He asked them to write one solution and explain the reasons for taking their approach, making sure not to criticize or make them feel they had made a mistake. Sometimes he took the problem in a new direction and asked the candidate to solve it a different way.

"When you take this approach, you see if the candidates have team player skills, whether they work well with you, and whether they go about solving a problem in a logical, sensible manner. I think the most important ability is to be able to think through the problem."

Aaron Underwood, an object expert with Andersen Consulting who manages Andersen's EAGLE advanced development group, judges technical know-how by asking candidates to discuss recent or current projects. He doesn't focus on whether or not someone knows a technology, but whether or not that person knows how to use it. "You get people who understand technology for the sake of technology, but who don't have a lot of experience in some of the fundamentals of development, such as life cycle methodology, testing models and such." When Underwood interviews potential staff members, he asks the candidates to talk about the challenges and how they've solved them. He asks them to discuss performance issues: What causes poor performance, what causes good performance?

"I probe for whether or not people have an awareness of how their work is handled when it's done. What makes their applications easy to support, what makes them not so easy to support. I'm looking to avoid the situation where you have someone whose quest for technical elegance isn't balanced with a dose of reality." Underwood sometimes poses the following:

- Give me an example of the worst production bug you experienced in a system you worked on.
- What would you have done in retrospect to have prevented that problem from happening?
- What's the best idea and worst idea you think you've put into a system, from the user's perspective?

To make sure the candidate is telling the truth about his or her experience, Underwood asks for details. "If somebody can't quote specifics to me about a situation, then they weren't there."

When it comes to specific technology areas, it's important to understand what you need in terms of the new technology before you write your interview questions. "There is no such thing as a universal set of questions that will smoke out a guru," says *Software Development* columnist, Roland Racko. "The first step in formulating your interview questions is to decide what properties have a high priority."

Making Your Decision

Underwood rates the candidates with respect to their knowledge and skill levels in system building fundamentals, development life cycle or methodology, and application operations and support. He balances these skills against their ability to learn new technologies.

Wang holds a post-interview meeting after a candidate's second interview, in which everyone who has interviewed the candidate expresses their views uninterrupted. Then, the whole group votes on whether or not to hire the person. "This way, everybody gets to articulate their concerns," explains Wang. " Listening to everybody speak up about what they're looking for helps tremendously in setting the expectations for the position within the team."

Each company will find its own path to an interviewing process that works, but some key points should be standard no matter what: Understand the position you need to fill before you try to fill it; get other people involved in the process; use tests and problem-solving not just to verify an applicant's skills, but also to discover how that person approaches his or her work and how he or she interacts with members of your team. Keep in mind that a well-defined interviewing process not only makes your job easier, it shows your prospective new employees that you care enough about your work to take them — and their skills — seriously.

Hiring Smart

Look for the ineffable spark of intelligence.

Watch for the play of emotion and intellect on the candidate's face. Is anybody home? You have to slow down a bit and analyze things to determine how deep the person is, how much potential the person has. Does the candidate challenge you to think more deeply, probe more subtly, assess more rapidly? Does the candidate draw out the best you have to offer? How does the whole encounter feel? Are you learning anything about anything? If the candidate is smart, his or her intelligence will probably have an impact on you and on the evolution of the interview itself. If the candidate can influence the course of the interview, it's likely that he or she can influence the course of other, more complicated events.

Challenge the candidate with a puzzle.

Naturally, you want the candidate to demonstrate the he or she will take reasonable approaches to solving the problem you pose, but more interesting are the observations you can make and the inferences you can draw about the candidate's working style. Does the candidate quickly make erroneous assumptions about the problem? Or does the candidate go to work on a clarification of the problem, soliciting more and more information until the problem domain has been clearly defined and how an answer might look has been fully envisioned?

Try teaching.

Assert your expertise, and see what happens. Does the candidate absorb what you have to offer? Ask for more information? Elaborate on your teaching? Apply it? Resist it? Does the candidate have a better idea?

A candidate's relationship to authority figures — to you in the interview situation — will be a key determinant of his or her future performance. You want to find out whether the candidate can set ego aside in order to solve a problem or learn something.

Take off the blinders.

The biggest mistake I see managers make as they hire people for software development teams is that they overvalue a particular technical skill. Obviously, conversance with a given technology is a wonderful attribute in a candidate, but in the final analysis it's an extra, not a mandatory.

— Excerpted from Jim McCarthy's *Dynamics of Software Development* (Microsoft Press, 1995).

Questions to Ask When Seeking the Object-Orientation Expert

- What kept you awake at night about the object technology side of the last project you worked on?
- How did you solve that problem?
- Who are Ivar Jacobson, Grady Booch, and Bertrand Meyer?
- What prevents a company from getting reuse to happen besides not using objects?
- What resources would you want in place before you'd promise to deliver reusable code?

— Roland Racko, Object Lessons columnist, *Software Development* magazine

Questions for Potential Database Programmers

- In a sub query, when would you use an IN vs. EXISTS?
- What's the difference between a primary key and a unique constraint?
- What's the difference between DRIP, DELETE and TRUNCATE?
- What is an outer join?
- What happens if you have a where clause that has a function such as substr on an indexed field?
- What's the difference between a function and a procedure?
- What's the difference between CHAR and VARCHAR?

— Mary Levine, senior information technology manager, Charles Schwab

Managers' Favorite Interview Questions

- How do you see the information systems role with respect to the business purpose? (If they don't know the answer to that question, they're just a techno jockey.)
- What steps do you go through before you roll out an application? Do you do thorough testing? Do you involve users? (This tells me whether or not they're careful.)
- How do you measure your own performance? What do you do at the end of the week or end of the month that helps you do your job better? Do you record and correct mistakes you make?(This gives me an idea of how motivated they are.)
- Do you think projects succeed or fail due to technical issues or user and management issues? (I want the answer to be user and management issues. You can always work around technical problems.)

— Amy Johnson, manager of corporate applications development, Miller Freeman Inc.

- What is the greatest technical achievement that you're proud to have been associated with?
- What unique algorithm have you developed or invented?
- What is your idea of the ideal project?

— Gene Wang, former vice president with Symantec and Borland, current CEO of Computer Motion

- Devise an algorithm to decode roman numerals. ("Hardly anybody gets this one, but when somebody does, I say O.K., this person knows how to program.")
— David Tenera, director of technology, Tenera Corp.

- What do you do when someone wants to build a new system? Do you use a particular methodology to help you? (I'm looking for things like needs analysis, requirements, design, development, implementation, documentation, and so on.)
- What kinds of standards exist in your current and/or previous jobs? Have you participated in defining them? (I am looking for the degree of structure a person is used to, his or her level of comfort with that structure, and a desire to contribute to standards.)
- A production system you are unfamiliar with has a serious problem that stops it from running. What do you do next? (I am looking for knowledge of who to ask, what to look for, how to approach solving a problem, and what sort of environmental expectations and experience a person has.)
— Mary Levine, information technology manager, Charles Schwab

- If you or the candidate is uptight during the interview, say something like, "This is just a script, and I'm acting like an interviewer. I'm going to drop that and try to act like a person. I'd like you to do the same."
— Jim McCarthy, author, *Dynamics of Software Development* (Microsoft Press, 1995)

For more interview questions and the reasoning behind them, see the *Software Development* web site at `http://www.sdmagazine.com`.

2.5.5 "Constant, Unceasing Improvement"
by Steve Maguire, January 1995

During the 1994 Winter Olympic Games, I was struck by one aspect of the figure skating events. The television footage of earlier gold medal performances seemed to suggest that 25 years ago you could win a gold medal with a few lay-back- and sit-spins, a couple of double toe loops, and a clean, graceful program. Today such a simple performance, however pleasing to watch, wouldn't win a hometown skating championship. Nowadays, you must do at least three triple jumps, several combination jumps, a host of spins, and lots of fancy footwork. On top of that, your program must have style, or the scores for artistic impression will look more like grade point averages than the 5.8s and 5.9s you need to win the gold.

At one point in the TV coverage, the commentator mentioned that Katarina Witt planned to skate the same program with which she had won the gold medal six years earlier at the Calgary Olympics. He added that it was unlikely Ms. Witt would place near the top even if she gave a clean performance — the very best programs only six years ago simply weren't demanding enough for competition today.

Think about that. Are skaters today actually better than the skaters of a quarter century ago? Of course, but not because Homo sapiens has evolved to a higher state of athletic capability. Some of the improvements in today's performances, I'm sure, are a result of better skates and ice arenas. But the dominant reason for the improvement is that each year skaters raise their standards as they try to dethrone the latest national or world champion. Skaters 25 years ago could have included all those triple and combination jumps in their routines, but they didn't need to, so they didn't stretch themselves to master those feats.

In the book *Peopleware* (Dorset House Publishing Co., 1987), Tom DeMarco and Timothy Lister describe a similar difference in standards of performance among programmers who work for different companies. DeMarco and Lister conducted "coding wars" in which they gave a programming task to two programmers from every company participating in one of the contests. They found that the results differed remarkably from one programmer to the next, at times by as much as 11 to 1 in performance. That disparity is probably not too surprising. The surprising news is that programmers from the same company tended to produce similar results. If one did poorly, so would the other. Likewise, if one did well, both did well, even though the two programmers were working independently. DeMarco and Lister point out that the work environments at the companies could account for some of the difference in performance among the companies, but I believe the major reason for the 11 to 1 variance is that the acceptable skill level for the "average programmer" varies from one company to the next.

When was the last time you heard a lead say to a programmer, "I'm disappointed in you. You're doing just what you're expected to do."? Whether a company is aware of the phenomenon or not, its programmers have an average skill level, and once a programmer reaches that average level, the pressure to continue learning eases up even though the programmer might still be capable of dramatic improvement. The programmers are like those ice skaters 25 years ago — good enough. And leads tend not to spend time training people who are already doing their job at an acceptable level. They work with people who haven't yet reached that level.

Having a team of programmers who do what is expected is not good enough. An effective lead perpetually raises the standards, as coaches for Olympic-class skaters must do. As you raise the programming standards of your team, you'll ultimately raise the standards — the average — of your whole company.

Five-Year Tenderfeet

Occasionally I'll run across a programmer who after five or more years still works on the same project he or she was first assigned to. No problem with that, but in many cases I find that the programmer is not only on the same project but also doing the same job. If the programmer was assigned to the Microsoft Excel project to work on Macintosh-specific features, for instance, that's what he'll still be doing — as the specialist in that area. If the programmer was assigned to the compiler's code optimizer project, years later she'll still be working on that isolated chunk of code — again, as the specialist.

From a project standpoint, creating long-term specialists for specific parts of your product is a good idea, but creating specialists can backfire if you don't educate them wisely. You'll cripple those programmers and ultimately hurt your project and your company if you don't see to it that your specialists continue to learn new skills.

Suppose that Wilbur, a newly hired programmer, spends his first year becoming your file converter specialist and then spends the next four years writing filters to read and write the file formats of competing products. There's no question that such work is important, but Wilbur will have gained a year's worth of real experience and then tapered off, learning little else for the next four years. Wilbur would claim that he has five years of programming experience, but that would be misleading — he would in fact have one year's experience five times over.

If Wilbur had spent the last four of those five years working on other areas of the application, he'd have a much wider range of skills. If he had been moved around to work on different aspects of a mainstream Windows or Macintosh application, for instance, he would have had an opportunity to develop all of this additional know-how:

- How to create and manipulate the user interface libraries — the menu manager, the dialog manager, the window manager — and all of the user interface gadgets you'd create with those libraries.

- How to hook into the help library to provide context-sensitive help for any new dialogs or other user interface extensions he incorporates into the application.

- How to use the graphics library to draw shapes, plot bit maps, do off-screen drawing, handle color palettes, and so on, for display devices with various characteristics.

- How to send output to printers, achieving the highest quality for each device, and how to make use of special features unique to each device, such as the ability of PostScript printers to print watermarks and hairlines.

- How to handle international issues such as double-byte characters, country- specific time and date formats, text orientation, and so on.

- How to handle the issues related to running an application in a networked environment.

- How to share data with other applications, whether the task is as simple as putting the data on the system clip-board or as complex as using CORBA or DCOM.

- How to write code that will work on all the popular operating systems — Windows, Linux, the Macintosh, and the Java virtual machine.

You get the idea. These skills are easily within the grasp of any programmer who works on a Windows or Macintosh application for five years — provided that every new task contains an as-yet-unlearned element that forces a programmer to learn and grow.

Compare the two skill sets. If you were to start a new team, which Wilbur would you want more, the five-year file converter specialist or the Wilbur with one year's experience in writing file converters plus four more years' experience with the varied skills in the list? Remember, both Wilburs have worked for five years.

A lead's natural tendency when assigning tasks would be to give all the file converter work to Wilbur because he's the specialist in that area. It's not until the Wilburs of the world threaten to leave their projects for more interesting work that leads switch mental gears and start throwing new and different tasks their way.

But "if the specialists aren't doing the tasks they're expert in, wouldn't they be working more slowly on tasks they know less about?" Or to put it another way, "Don't you lose time by not putting the most experienced programmer on each task?"

If you view the project in terms of its specific tasks, the answer must be, "Yes, each task is being done more slowly than it could be done by a specialist." However, that little setback is more than compensated for when you look at the project as a whole. If you're constantly training team members so that they're proficient in all areas of your project, you build a much stronger team, one in which most team members can handle any unexpected problem. If a killer bug shows up, you don't need to rely on your specialist to fix it — anybody can fix it. If you need to implement a new feature in an existing body of code, any of many team members can efficiently do the work, not just one. Team members also know more about common subsystems, so you reduce duplicate code and improve product-wide design. The entire team has versatile skill sets.

Your team may be losing little bits of time during development as they learn new skills and gain experience, but for each minute they lose learning a new skill, they save multiple minutes in the future as they use that skill again and again. Constant training is an investment, one that offers tremendous leverage and tremendous rewards.

Don't allow programmers to stagnate. Constantly expose each team member to new areas of the project.

Reusable Skills

At Microsoft, when a novice programmer moves onto a project, he or she is typically given introductory work such as tracking down bugs and incorporating small changes here and there. Then gradually, as the programmer learns more about the program, the tasks become increasingly more difficult, until the programmer is implementing full-blown mega-features. This gradualist approach makes sense because you can't very well have novices making major changes to code they know nothing about. My only disagreement with this approach is that the tasks are assigned according to their difficulty rather than according to the breadth of skills they could teach the programmer. As you assign tasks to programmers, keep the skills-teaching idea in mind. Don't assign successive tasks solely on the basis of difficulty; make sure that each task will teach a new skill as well, even if that means moving a novice programmer more quickly to difficult features. Even better, assign tasks at first that teach skills of benefit not only to your project but to the whole company.

In a spreadsheet program, for instance, tasks might range from implementing a new dialog of some sort to working on the recalculation engine. The skills a programmer would learn from these two tasks fall at two extremes: one skill has nothing to do with spreadsheets specifically, and the other historically has little use outside spreadsheet programming. Putting a programmer on the recalculation engine would be educational and would provide a valuable service to the project, but the skill wouldn't be as transferable as knowing how to implement a dialog would be. Learning how to create and manipulate dialogs could be useful in every project the company might undertake.

Creating a better "average programmer" means raising the standard through-out the company, not just on your project. You could assign programmers a random variety of tasks and ensure that team members would constantly learn, but you can do better than that. Analyze each task from the standpoint of the skills it calls upon, and assign it to the programmer who most needs to learn those skills. An experienced programmer should already know how to create dialogs, manipulate windows, change font sizes, and so on. She is ready to develop less globally useful — more specialized — skills such as the ability to add new macro functions to the spreadsheet's macro language. At some point, she'll know the program so well that in order to continue learning she'll have to move to extremely project-specific work such as implementing an even smarter recalculation engine.

A novice team member should be assigned a few tasks in which he must learn to create dialogs, followed by a few tasks that force him to manipulate windows, and so on. Deliberately assign tasks that cumulatively require all the general skills. That way, if the division should be reorganized and the programmer should find himself on another project, the skills he's learned will still be useful.

This is another example of a small system that produces greater results. Which specific work you assign to a novice programmer may not make much difference in the progress of your own project, but by first exposing a new programmer to a wide range of general skills that he or she can bring to any project, you make the programmer more valuable to the company.

When training programmers, focus first on skills that are useful to the entire company and second on skills specific to your project.

The Cross-Pollination Theory Dismissed

Occasionally I'll run across the idea that companies should periodically shuffle programmers around so that they can transfer ideas from one project to another. It's the cross-pollination theory. The cross-pollination theory appeals to me because its purpose is to improve development processes within the company, but in my experience the cross-pollination practice falls short of its goal, and for a simple reason: it ignores human nature.

Advocates of the theory assume that programmers who move to brand-new groups will teach the new groups the special knowledge they bring with them. But how many people feel comfortable doing that in a new environment? And even if a programmer would feel comfortable as an evangelist, how many groups would appreciate some newcomer's telling them what they should do? A new lead might feel fine propounding fresh ideas hours or days into the project, but nonleads? It might be years, if ever, before a programmer would feel comfortable enough to push his or her ideas beyond a narrow work focus.

Advocates of the cross-pollination theory assume that new people bring new knowledge into the group. In fact, that's backwards from what actually happens: new people don't bring their knowledge into the new group as much as they get knowledge from the new group. New people find themselves immersed in different, and possibly better, ways of doing things. And they learn. The primary benefit is to the person doing the moving. If that person can continue to grow on his or her current project, why cause disruption? Let the people who are stagnating move to other teams and learn more. Don't shuffle people around to other teams expecting them to spread the word. They usually won't.

Give Experts the Boot

If you constantly expose a team member to new tasks that call for new skills, he or she will eventually reach a point at which your project no longer offers room to grow. You could let the programmer's growth stall while your project benefited from his or her expertise, but for the benefit of the company, you should kick such an expert off your team. If you allow programmers to stagnate, you hurt the over-all skill level of the company. You have a duty to the programmers and to the company to find the programmers positions in which they can grow.

Am I joking? No.

The tendency is to jealously hold onto the team's best programmers even if they aren't learning anything new. Why would you want to kick your best programmer off the team? That would be insane.

In Chapter 3 of *Debugging the Development Process*, I talked about a dialog manager library that the Word for Windows group had been complaining about. Although I wasn't the lead of the dialog manager team then, I did eventually wind up in that position. And there came a point at which the main programmer on the team had reached a plateau: he wasn't learning anything new within the constraints of that project. Besides, he was tired of working on the same old code. He needed to stretch his skills.

When I asked whether he knew of any interesting openings on other projects, he described a position in Microsoft's new user interface lab in which he would be able to design and implement experimental user interface ideas. In many ways, it seemed like a dream job for the program-mer, so I talked to the lab's director to verify that the job was a good opportunity for this programmer to learn new skills. The position looked great. In less than a week, the dialog team's best programmer was gone, leaving a gaping hole.

In these situations, you can either panic or get excited. I get excited because I believe that gaping holes attract team members who are ready to grow and fill them. Somebody always rises to the occasion, experiencing tremendous growth as he or she fills the gap. The dialog team's gap proved to be no different. Another member jumped headlong into the opening.

Occasionally I'd bump into the lab director and ask how the project was going. "Beyond my wildest dreams," he'd say. "We're accomplishing more than I ever imagined or hoped for." He had been expecting to get an entry-level programmer, but he'd gotten a far more experienced programmer, and his group was barreling along.

The dialog manager group with its new lead programmer was barreling along too. The new lead had just needed the room to grow, room that had been taken up by the expert pro-grammer.

You might think that kicking your best programmer off the team would do irreparable harm to your project. It rarely works out that way. In this case, the dialog team experienced a short-term loss, but the company saw a huge long-term gain. Instead of a slow-moving user inter-face project and two programmers who had stopped growing, the company got a fast-moving user interface project and two programmers who were undergoing rapid growth. That outcome shouldn't be too surprising. As long as its people are growing, so is the company.

Don't jealously hold onto your best programmers if they've stopped growing. For the good of the programmers, their replacements, and the company, transfer stalled programmers to new projects where growth can continue.

The New Year's Syndrome

Not all skills can be attained in the course of doing well-chosen project tasks. A skill such as learning to lead projects must be deliberately pursued as a goal in itself. The person must decide to be a good lead and then take steps to make it happen. It's proactive learning, as opposed to learning as a side effect of working on a task.

If you want your team members to make great leaps as well as take incremental daily steps to improvement, you must see that they actively pursue the greater goals.

The traditional approach to establishing such goals is to list them as personal skill objectives on the annual performance review. We all know what hap-pens to those goals: except for a few self-motivated and driven individuals, people forget them before the week is over. Then along comes the next review, and their leads are dismayed to see that none of the personal growth goals has been fulfilled. I think we've all seen this happen — it's the New Year's Resolution Syndrome, only the date is different.

Such goals fall by the wayside because there are no attack plans for achieving them or because, if there are such plans, the plans have no teeth — like those postmortem plans I spoke of in Chapter 4 of *Debugging the Development Process* had no teeth. Listing a goal on a review form with no provision for how it will be achieved is like saying, "I'm going to be rich" but never deciding exactly how you're going to make that happen. To achieve the goal, you need a concrete plan, a realistic deadline, and a constant focus on the goal.

One way to ensure that each team member makes a handful of growth leaps each year is to align the personal growth goals with the two-month project mile-stones. One goal per milestone. That practice enables team members to make six leaps a year — more if there are multiple goals per milestone.

Improvement goals don't need to be all-encompassing. They can be as simple as reading one good technical or business book each milestone or developing a good habit such as stepping through all new code in the debugger to proactively look for bugs. Sometimes the growth goal can be to correct a bad work habit such as writing code on the fly at the keyboard — the design-as-you-go approach to programming.

To ensure their personal interest in achieving such goals, I encourage team members to choose the skills they want to pursue, and I merely verify that each goal is worth going after if it meets the following conditions:

- The skill or knowledge would benefit the programmer, the project, and the company. Learning LISP could be useful to an individual, but for a company such as Microsoft, it would be as useful as scuba gear to a swordfish.

- The goal is achievable within a reasonable time frame such as the two-month milestone interval. Anybody can read a good technical book in two months. It's much harder to become a C++ expert in that short a time.

- The goal has measurable results. A goal such as "becoming a better programmer" is hard to measure, whereas a goal such as "developing the habit of stepping through all new code in the debugger to catch bugs" is easy to measure: the programmer either has or hasn't developed the habit.

- Ideally, the skill or knowledge will have immediate usefulness to the project. A programmer might acquire a worth-while skill, but if he has no immediate use for the new skill, he's likely to lose or forget what he's learned.

Such a list keeps the focus on skills that are useful to the individual, to his or her project, and to the company — in sum, it focuses on the kinds of skills a programmer needs in order to be considered for promotion. If the programmer can't think of a skill to focus on, choose one yourself: "What additional skills would this programmer need for me to feel comfortable about promoting him or her?"

Make sure each team member learns one new significant skill at least every two months.

Read Any Good Books Lately?

Reading constantly is something I do to gain new knowledge and insights. Why spend years of learning by trial and error when I can pick up a good book and in a few days achieve insights that took someone else decades to formulate? What a deal. If team members read just six insightful books a year, imagine how that could influence their work.

I particularly like books that transform insights into strategies you can immediately carry out. That's why I wrote *Writing Solid Code* (Microsoft Press, 1993) and this book as strategy books. But mine are hardly the first. *The Elements of Programming Style*, by Brian Kernighan and P. J. Plauger (Yourdon/McGraw Hill) was published in 1974 and is still valuable today. *Writing Efficient Programs* (Prentice Hall, 1982) by Jon Bentley, is another excellent strategy book, as is Andrew Koenig's *C Traps & Pitfalls for C and C++ Programmers* (Addison Wesley, 1989).

In addition to these strategy books, there are dozens of other excellent — and practical — books on software development, from Gerald Weinberg's classic *The Psychology of Computer Programming* (V. Nostrand Reinhold, 1971) to the much more recent *Code Complete* (Microsoft Press, 1993) by Steve McConnell, which includes a full chapter on "Where to Go for More Information," with brief descriptions of dozens of the industry's best books, articles, and organizations.

But don't limit yourself to books and articles that talk strictly about software development. Mark McCormack's *What They Don't Teach You at Harvard Business School* (Bantam, 1986), for instance, may focus on project management at IMG, his sports marketing firm, and Michael Gerber's *The E-Myth* (Harpers, 1985) may focus on how to build franchise operations, but books like these provide a wealth of information you can apply immediately to software development. And don't make the mistake of thinking that such books are suit-able only for project leads. The greenest member of the team can benefit from such books.

In the Moment

A particularly good approach to identifying skills for your team members to develop is to set a growth goal the moment you see a problem or an opportunity. When I spot programmers debugging ineffectively, I show them a better way and get them to commit to mastering the new practice over the next few weeks. When a programmer remarks that she wants to learn techniques for writing fast code, I hand her a copy of Jon Bentley's Writing Efficient Programs and secure her commitment to reading it — and later discussing it. If I turn up an error-prone coding practice as I review some new code, I stop and describe my concern to the programmer and get him to commit to weeding the practice out of his programming style.

I'm big on setting improvement goals in the moment. Such goals have impact because they contain a strong emotional element. Which do you think would influence a programmer more: showing him code he wrote a year ago and asking him to weed out a risky coding practice or showing him a piece of code he wrote yesterday and asking him to weed out the practice?

I once trained a lead who would search me out every time he had a problem. He'd say, "The WordSmasher group doesn't have time to implement their Anagram feature, and they want to know if we can help out. What should we do?" The lead came to me so often that I eventually concluded he wasn't doing his own thinking. When I explained my feelings to him, he replied that he always thought through the possible solutions but didn't want to make a mistake. That was why he was asking me what to do. I pointed out that his approach made him seem too dependent and that we needed to work on the problem.

I understood the lead's need for confirmation, so I told him to feel free to talk to me about problems as they arose, on one condition. Instead of dumping the problem in my lap, he was to:

- Explain the problem to me.
- Describe any solutions he could come up with, including the pros and cons of each one.
- Suggest a course of action and tell me why he chose that course.

Once the lead began following this practice, my perception of him changed immediately and radically. On 9 out of 10 occasions, all I had to do was say, "Yes! Do it," to a fully considered plan of action. The few times I thought a different course of action made sense, I explained my rationale to him, we talked it over, and he got new insights. Sometimes I got the new insights. We'd go with his original suggestion if my solution was merely different and not demonstrably better.

This improvement took almost no new effort on either his part or mine, but the shift in his effectiveness was dramatic. We went from a relationship in which I felt as if I were making all his decisions to one in which I was acknowledging his own good decisions. My attitude changed from "this guy is too dependent and doesn't think things through" to "this guy is thoughtful and makes good decisions." His attitude changed too, from being afraid to make decisions to knowing that most of his decisions were solid. It didn't take too many weeks for our "What should I do?" meetings to all but disappear. He consulted me only for truly puzzling problems for which he couldn't come up with any good solution.

What caused this dramatic change? Was it a major revamping of this person's skills? No, it was a simple change in communication style provoked by my realization that he had become too dependent. A minor change, a major improvement.

Take immediate corrective action the moment you realize that an area needs improvement.

Train Your Replacement

Programmers don't usually choose to pursue management skills unless they have reason to believe they're going to need those skills. Find the people who have an interest in becoming team leads, and help them acquire the skills they'll need to lead teams in the future. And remember, unless you plan to lead your current team forever, you need to train somebody to replace you. If you don't, you might find yourself in a tough spot, wanting to lead an exciting new project and unable to make the move because nobody is capable of taking over your current job.

After-the-Fact Management

Note that I gave that lead on-the-spot feedback and a goal he could act on immediately. I didn't wait for the annual review. I don't believe the annual review is a good tool for planning personal improvement or achievement goals. In my experience such a delayed response to problems isn't effective — at least not unless the annual review also contains detailed attack plans for the goals. Another problem with using the annual review for improvement goals is that few leads are able to effectively evaluate anyone's growth over such a long period of time.

We've all heard stories about the review in which the manager brings up a problem with the programmer's performance that has never been mentioned before to justify giving the programmer a review rating lower than the programmer expected. In shock, the programmer stammers, "Can you give me an example of what you're talking about?" The manager stumbles a bit and comes up with some-thing, that, yes, the programmer did do, or failed to do, but which sounds absurdly out of proportion in the context of the programmer's performance for the whole review period. "You've given me a low rating because of that?" Of course, it sounds ridiculous to the manager too, so she scrambles to come up with another example of the problem but usually can't because so much time has passed.

Then, of course, once the programmer leaves the meeting and has time to think about the review a bit, his or her reaction is anger. "Why didn't she tell me something was wrong, rather than waiting a year? How could I have fixed something I didn't even know was wrong?"

I've lost track of the number of times I've heard people say that about their managers.

What if professional football teams worked that way? What if coaches waited until the end of the season to tell players what they're doing wrong?

"Mad Dog, I'm putting you on the bench next season."

"Huh? What? I thought I played great," says Mad Dog, confused.

"You played well, but at each snap of the ball, you hesitated before running into position."

"I did?"

"Yes, you did, and that prevented you from catching as many passes as you could have. I'm putting you on the bench until something changes. Of course, this means that your yearly salary will drop from $5.2 million to $18,274. But don't worry, you'll still have your benefits — free soft drinks and hot dogs at the concession stand, and discounted souvenirs."

Mad Dog, particularly mad now: "If you spotted this, why didn't you tell me earlier? I could have done something about it."

"Hey, I'm telling you now, at our end-of-the-season contract review."

Sounds pretty silly, doesn't it? But how does it differ from the way many leads make use of the annual review?

Remember the lead I felt was too dependent and was not thinking things through? The common approach at most companies would be to wait until the end of the review period and note the problem on the review document:

"Relies too much on other people to make his decisions; doesn't take the time to think problems through."

Then, of course, after the confused exchange at the review meeting, the attack plan to fix the problem would be some-thing like this:

"I won't rely on other people to make my decisions for me; I'll think my problems through."

That attack plan won't be effective because it is too vague. The plan doesn't describe what the person is to do, how he is to do it, or how to measure the results — the plan has no teeth. In all likelihood, the problem will still exist a year later, at the next review.

Personnel reviews, as I've seen them done, are almost totally worthless as a tool to promote employee growth. Don't bother with the new goals part of the review. Actively promote improvement by seizing the moment and aligning growth goals with your project milestones. Use the for-mal review to document employee growth during the review period — that's what upper management really needs to see anyway. Listing areas in which people could improve doesn't really tell upper management much. Documenting the important skills that people have actually mastered and how they applied those skills demonstrates constant growth and gives upper management something tangible with which to justify raises, bonuses, and promotions.

Don't use the annual personnel review to set achievement goals. Use the review to document the personal growth goals achieved during the review period.

Tips for Constant Team Improvement

- Never allow a team member to stagnate by limiting him or her to work on one specific part of your project. Once programmers have mastered an area, move them to a new area where they can continue to improve their skills.

- Skills vary in usefulness from those that can be applied to any project to those that can be applied to only one specific type of project. When you train your team members, maximize their value to the company by first training them in the most widely useful skills and save the project-specific skills for last.

- It's tempting to hold onto your top programmers, but if they aren't learning anything new on your project, you're stalling their growth and holding the company's average skill level down. When a top programmer leaves the team for a new position, not only does he or she start growing again, but so does his or her replacement. A double benefit.

- To ensure that the skills of the team members are expanding on a regular basis, see that every team member is always working on at least one major improvement goal. The easiest approach is to align growth goals with the two-month milestones, enabling at least six skill leaps a year — which is six more per year than many programmers currently experience. If Wilbur, the file converter specialist, had read just 6 high-quality technical books a year, after his first five years of programming he'd have read 30 such books.

How do you suppose that would have influenced his work? Or what if Wilbur had mixed the reading of 15 good books with the mastery of 15 valuable skills over that first five years?

- The best growth goals emerge from a strong, immediate need. If you find a team member working inefficiently or repeating the same type of mistake, seize the opportunity to create a specific improvement goal that the team member can act on immediately. Because such on-the-spot goals lend themselves to immediate action for a definite purpose, the programmer is likely to give them more attention than he would give to abstract goals devised for an annual review.

Thoroughly Knowledgeable

Most of the interviews I conducted at Microsoft were with college students about to graduate, but occasionally I interviewed a working programmer who wanted to join Microsoft. At first I was surprised to find that the experienced programmers who came from small, upstart companies seemed, in general, more skilled than the experienced programmers from the big-name software houses, even though the programmers had been working for comparable numbers of years. I believe that what I've been talking about in this chapter accounts for the difference. The programmers working for the upstart companies had to be knowledgeable in dozens of areas, not expert in one. Their companies didn't have the luxury of staffing 30-person teams in which each individual could focus on one primary area. Out of necessity, those programmers were forced to learn more skills.

As a lead — even in a big outfit that can afford specialists — you must create the pressure to learn new skills. It doesn't matter whether you teach team members personally or whether they get their training through books and technical seminars. As long as your teams continue to experience constant, unceasing improvement, the "average programmer" in your company will continue to get better — like those Olympic-class skaters — and that can only be good for your project, for your company, and ultimately for your customers.

2.5.6 "Object-Oriented Training"

by Scott W. Ambler, September 1996

Orient your developers toward objects by addressing their individual needs.

Languages like Java, C++, and potentially ObjectCobol dominate the development landscape, and all of them are based on the object-oriented paradigm. Before you and your team can work in these environments, you must learn object-oriented development.

Object orientation is a whole new development ball game — one you can't pick up overnight. It takes between six and nine months for a developer with several years of experience to gain a working knowledge of object orientation, and potentially another year or two to truly understand the paradigm. Worse, the more experience a person has in structured development, the harder it is for that person to learn object orientation. He or she will have more to unlearn than someone with less experience.

Despite these challenges, object orientation is here to stay and the sooner you get your team trained in this discipline the better. Here are a few ways to get them going.

Formal Training Courses

Formal courses are the most common approach to teaching people object orientation. Although courses can be beneficial, most people can't learn all they need to know from taking only one. People typically need a series of courses to become effective object-oriented developers.

In Figure 2.3, we see four suggested training plans: one for programmers, one for analysts and designers, one for project managers, and one for executives. They all start with courses that cover the fundamentals. Don't kill the learning process by forcing people to start with things they aren't interested in. For the most part, programmers learn best when programming courses are among the first things they do, whereas analysts and designers learn best when they start out with analysis and design courses.

Everyone needs training in the object-oriented system development life cycle. A one-day overview of the object-oriented development process is critical to gain a basic understanding of how the life cycle works and why it is different from the structured approach. Because programmers are involved in the real work of application development, they need courses on specific aspects of object-oriented development.

Figure 2.3 shows the recommended periods between courses to give people time to internalize the things they learn. Some "object-oriented university" programs — where developers spend two or three months off site in a combination of training classes and mentored development environments designed to speed up the learning process — don't provide this downtime. While designed to speed up the learning process, these programs rarely live up to their full potential.

Figure 2.3   Recommended periods between courses.

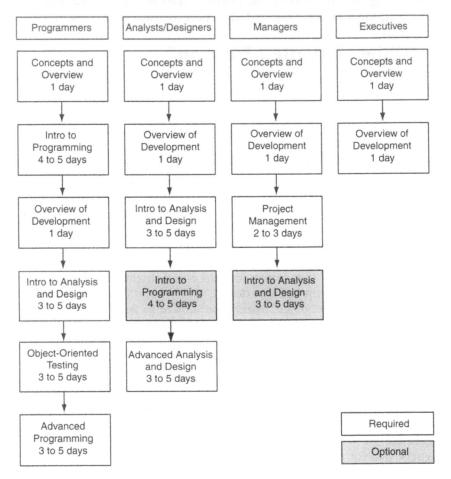

Enhancing the Training Process

"Just-in-time training" is important. Give your people training when they need it, not several months before or after. Studies show that within 21 days of training, people will forget 75% of what they learned unless they apply their new skills immediately after the course. In other words — use it or lose it.

When putting together a training and education program in object orientation, you may also need to teach skills in related topics such as GUI design, client/server development, and personal computer use. Very often the move to object-oriented development is precipitated by a move to a new hardware or software platform. Moving to a new environment means you'll need to give your staff the skills to work in and develop for it.

Your organization should also provide a wide range of books, magazines, and journals on object-orientation to everyone being trained. People should be given an opportunity to do some reading before they go on a course so that they can come prepared to ask good questions. It's amazing how much a $35 book can enhance a $2,000 training course.

Learning Teams

An effective way to train people in object orientation is to put them into learning teams. These groups work together on a small object-oriented project, perhaps something for human resources or the marketing department. They usually spend between 20% and 50% of their working hours on the project.

The best learning teams include people from different areas in your systems department who have various skills. Perhaps one is a manager, another a systems programmer, another an analyst, and so on. This wide range of skills and backgrounds enables the team to approach the learning process from several directions, thus increasing learning opportunities. Team members should focus on learning object-orientated concepts like inheritance, polymorphism, and encapsulation. They'll also need access to object-oriented development tools and literature.

Learning teams have a tendency to flounder without the guidance of an experienced mentor. Good mentors are hard to find. They should have at least two or three years of solid object-oriented development experience, have been involved in the entire development process on several projects, good communication skills, and they should care about the people they are helping. The mentor ensures the students build good object-oriented skills and prevents them from picking up bad habits.

The mentoring process typically takes between six to twelve months, with the mentor working a full-time schedule at the beginning of a project and a day or two per week toward the end. The trick is to slowly wean yourself off your mentors by having them transfer their skills to your staff. Good mentors make you independent of them, bad mentors don't.

Fending for Yourself

Developers often find themselves in organizations that either can't afford to, or simply won't, train their staff in object-oriented development. Here are few pointers for people learning on their own:

- Start reading. Pick up a couple of introductory books, a few magazines, and go at it. Your main goal is to get a feel for what object-orientation is all about and how to apply what you read throughout the entire development life cycle.

- Learn Smalltalk. One of the best ways to teach yourself object-orientation is to learn Smalltalk, a high-level, pure object-oriented language supported by vendors such as ParcPlace/Digitalk, IBM, and Cincom. Smalltalk is considered pure because you can only do object-oriented programming in it, unlike a hybrid language like C++. The main advantage of starting with Smalltalk is that it doesn't let you slip back into your old, evil, structured development ways. Check out the web page
  `http://sumeru.stanford.edu/learningworks` for software to teach yourself Smalltalk.

- Sign up for a night course at your local college. Look for instructors who have taught the course several times, and who have real-world object-oriented development experience.
- Get the word out. Make your manager aware that you are learning object-orientation at home and that you are keenly interested in getting on an object-oriented project. Hopefully you'll be put at the top of the list of people to get on the next project, which is exactly what you want. It doesn't do any good to learn the skills if you don't apply them.

Bag-Lunch Sessions

One-hour mini-lessons held during the daily lunch break are a great supplement to other training. The sessions are typically given by object-oriented development experts, usually your mentors, and cover a wide range of topics. One day the lesson may be about Smalltalk collection classes and the next day about managing an object-oriented development team. Successful bag-lunch training programs typically include two or three sessions per week, with each individual session provided several times so that everyone has an opportunity to attend.

Computer-Based Training

Computer-based training is also a valid approach, especially when it's combined with formal training and mentoring. Many organizations provide employees with access to introductory computer-based training courses before sending them to formal training courses. Unfortunately, computer-based training alone is of minimal value for teaching people object-oriented development.

Most developers learn by playing, reading, and then playing some more. This means they need object-oriented programming languages, modeling tools, and testing tools. You should also buy everyone several books, ranging from introductions, to programming, to analysis and design. Magazines can also go a long way so get a few subscriptions.

You don't need to train everyone in object orientation. You'll have to maintain your legacy applications for years to come. Some people don't want to be trained in object orientation, or are afraid to try, or else they truly enjoy maintaining Cobol code. Regardless of their motivation, if these people want to stay doing what they're doing, let them.

Second, contrary to popular belief, nobody ever quit because they were trained. Developers quit because the money isn't good enough, the work isn't interesting enough, or because they don't like the people they're working with. Object-oriented developers are paid more than other developers, and any organization that enters into object-oriented training and education had better be prepared to pay their people what they're worth. Object-oriented developers are in demand, so don't get scooped by your competition. Treat your people well.

Challenges of the Object-Oriented Education Process

- *Confronting the structured programming mentality.* Structured developers have difficulty determining when and when not to apply the new object-oriented concepts that they have learned, what concepts from the structured environment still apply, and which ones don't. These problems diminish with experience, but until a developer gets that experience he or she is going to have some growing pains. Experienced mentors and patient team leaders are critical to their success.

- *Not everyone learns the same way.* Some people learn best in the classroom, others by sitting down and working with a language, and others through working in teams. Because no training and education plan is perfect for everyone, you will want to create an approach that can be modified to meet individual needs.

- *Dealing with bruised egos.* Experienced developers learning objects go from being a recognized expert to a recognized novice. This hurts. Developers need to realize that if they apply themselves they can become experts once again, it just takes a while.

- *Dealing with the "I've done it before" syndrome.* 70% of object-orientation is based on some of the same principles that structured and procedural development is based on. It's really easy to fool yourself into thinking that you already understand object-orientation. It's that additional 30% that makes the difference. As soon as developers start to work on a real project with good mentors, they quickly realize there's a lot more to object-orientation than what they originally thought.

- *Dealing with the "It's just another fad" syndrome.* It isn't unusual to find people who think object-oriented development is just a fad. Sometimes this is a reflection of the fear of learning, and sometimes it's an actual belief. Try to get them to see the light. If people are either unable or unwilling to realize the importance of object-orientation, then perhaps you should move them to a project that isn't object-oriented.

- *Taking care of the people you aren't training.* On the first couple of object-oriented projects, many developers won't be involved. Keep them up to date about what's being learned, let them know when and how they'll be involved later, give them access to the tools on off-hours so they can learn on their own, invite them to bag-lunch training sessions, and give them access to object-oriented books and magazines. If you don't involve them you risk losing them, forcing yourself to move people from object-oriented projects to support legacy applications.

- *Realizing that some colleges and universities are not up to speed.* Object-orientation has caught many computer science departments off-guard — if they weren't actively involved in object-oriented research, chances are they haven't had a curriculum in place very long. Avoid courses that are being taught for the first time, as you run the risk of getting an instructor who is only a chapter or two ahead of the students.

- *Getting people into training quickly.* While self-study is good, it's too easy to misunderstand concepts. Professional instructors can help you learn object-orientation properly and avoid bad habits you might otherwise pick up.

Chapter 3

Best Practices for the Business Modeling Workflow

The purpose of the Business Modeling workflow is to model the business context of your system in the form of a business model. A business model typically contains:

- A context model that shows how your system fits into its overall environment. This model will depict the key organizational units that will work with your system, perhaps the marketing and accounting departments, and the external systems that it will interact with.

- A high-level use case model, typically a portion of your enterprise model (perhaps modeled in greater detail), that shows what behaviors you intend to support with your system.

- A glossary of key terms

- An optional high-level class diagram (often called a *Context Object Model* or *Domain Object Model*) that models the key business entities and the relationships between them.

- A business process model, traditionally called an *analysis data-flow diagram* in structured methodologies, depicts the main business processes, the roles/organizations involved in those processes, and the data flow between them. Business process models show how things get done, as opposed to a use case model that shows what should be done.

Your goal is to evolve a common understanding of your system with its stakeholders, including your users, senior management, user management, your project team, your organization's architecture team, and potentially even your operations and support management. Part of the consensus between your project stakeholders is agreement as to the scope of your project — what your project team intends to deliver. Without this common understanding, your project will likely be plagued with politics and infighting and could be cancelled prematurely if senior management loses faith in it. Throughout your project, it scope will be managed via the Configuration and Change Management workflow.

Your goal is to reach consensus between project stakeholders.

One of the very first steps of the business modeling process is to identify who your stakeholders are. Who will use your system to support their jobs? Who is the person(s), often senior management, paying for the development of your system? Who will operate your system once it is in production? Who will support your system once it is in production? Once you know who your stakeholders are, you need to invest the time to understand what it is that they want from the system — key input into your business modeling efforts.

You need to understand who your project stakeholders are.

Your business model should be driven by your enterprise model (a deliverable of the Infrastructure Management workflow described in Chapter 5) which models the high-level business requirements for your organization. Figure 3.1 depicts how the business model fits into the scheme of things; in many ways, Figure 3.1 is a context model. The figure shows that the enterprise model drives the development of the business model — the business model is effectively a more detailed view of a portion of your overall business. It isn't purely a top-down relationship, however, because as your business analysts learn more about the business problem that your system addresses, they will share their findings back to the owners of the enterprise model who will update it appropriately. The business model, in turn, drives the development of your requirements model. The Requirements workflow, covered in Chapter 4, similarly provides feedback for the business model.

Figure 3.1   How the business model fits in.

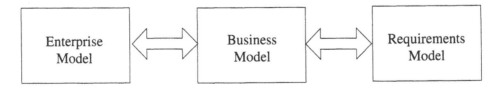

3.1 From Modeling Your Business to Business Process ReEngineering

Another goal that you may have for this workflow is to perform business process reengineering (Hammer & Champy, 1993; Champy, 1995), also know as *BPR*. The purpose of BPR is simple, it is "the fundamental rethinking and radical redesign of business processes to achieve dramatic improvements in critical, contemporary measures of performance, such as cost, quality, service, and speed." For example, Amazon.com effectively reengineered the concept of selling books by creating an online virtual bookstore. At their web site, anyone can search for books, order what they want, post reviews of books that they have read, and have the books that they order sent to them within days from the publisher's warehouse. Amazon typically doesn't own or even touch the books that you purchase, they merely act as a broker between you and the publisher. On the other hand, many of the most successful companies online are often existing physical businesses whose fundamental business processes have been reengineered to use effectively use modern technology, such as Dell computers selling its product line via the Internet.

Business process reengineering (BPR) involves the fundamental rethinking and radical redesign of business processes.

What isn't so simple is how to perform business process reengineering, which is why you need to apply a technique such as *essential modeling* so that you may first understand the true business problem to be addressed. In "Essentially Speaking," (*Software Development*, November 1994) Larry Constantine presents the concept of an essential model — a model that represents the core aspects of a problem stripped of all unnecessary assumptions. Essential models capture a technology-free, abstract view of a system, describing the fundamental features that the system offers to its users. In many ways, essential modeling is an idealized form of business modeling, although admittedly, it also overlaps into the Requirements workflow too. In this article, Constantine describes essential use case models — very similar conceptually to logical business process models from the structured world — that can be used to drive your organization's business process reengineering efforts. In this article, Constantine introduces the concept of *essential user interface models*, a topic that he describes in detail in "Prototyping From the User's Viewpoint" (*Software Development*, November 1998) which is included in Chapter 4. The techniques and concepts presented in this article form a solid collection of best practices from which you may conduct your business modeling and business process reengineering efforts.

Essential modeling is a fundamental skill for the Business Modeling workflow and is a key enabler of BPR.

To be fair, BPR has gotten a bad reputation the last couple of years due to several public reengineering failures. When you look into these failures, however, you quickly discover that the reengineering efforts were often poorly executed — it wasn't BPR that was at fault but instead the people who misapplied it. Common causes of failure typically included lack of commitment by senior management and lack of involvement of information technology (IT) staff. The appropriate application of information technology is often the backbone of any reengineering effort, supported by changes to your organizational culture and business processes. A proven software process, such as the enhanced version of the Unified Process described in this book series, increases the chance that you will apply information technology effectively in your BPR efforts.

BPR can be difficult to implement successfully. Following a proven and effective software process will increase your chances dramatically.

The identification of business rules is an important aspect of both the Business Modeling workflow and the Requirements workflow. In "The Logic of Business Rules" (*Software Development*, November 1997), Colleen McClintock describes the process for collecting and managing business rules as well as how to incorporate these processes in your overall modeling effort. According to McClintock, "a business rule is an atomic, explicit, and well-formed expression that describes or constrains the business principles, guidelines, and operations of a company using vocabulary and syntax that can be easily used and understood by the persons within the company who are responsible for defining and carrying out the business." With respect to the Business Modeling workflow, your goal is to identify and document the critical business rules that are pertinent to your project. During the Requirements workflow, you will evolve these business rules, and identify new and often more granular business rules, and incorporate them into your models. In a related article, "Engineering Object-Oriented Requirements" (*Software Development*, March 1999), I show how business rules fit into your overall modeling efforts — your use cases will refer to business rules and your components and classes will implement them. In a sidebar, McClintock provides insight into how to model business rules in UML diagrams using the object-constraint language (Warner and Kleppe, 1999), more commonly referred to as *OCL*. Although she is correct that you can model business rules in this fashion, my experience is that you also need prose descriptions in a format that your users can understand — descriptions that are likely to be part of your Supplementary Specifications artifact (Kruchten, 1999). Finally, McClintock provides excellent advice for implementing business rules using business rule servers and/or commercial rules engine.

Business rules must be taken into account at all levels of modeling.

3.2   A Common Modeling Notation

An important goal of the Business Modeling workflow is to engender a common understanding of the system that your project team will deliver. To achieve this goal you need to communicate and work with the stakeholders on your project and to communicate effectively, you need a common language. The Unified Modeling Language (UML) is the definition of a collection of diagrams — including use case diagrams, class diagrams, and activity diagrams — that your project team can use to model your system (Rumbaugh, Jacobson, & Booch, 1999). The UML describes a common notation, and the semantics for applying that notation, that you may apply for business modeling, enterprise modeling, requirements modeling, and so on. The UML is a standard that is supported and evolved by the Object Management Group (OMG), their web site is http://www.omg.org.

The UML provides a standard communication language for your development efforts.

In "Why Use the UML?" (*Software Development*, March 1998), Martin Fowler, author of *Analysis Patterns* (Fowler, 1997) and co-author of *UML Distilled* (Fowler and Scott, 1997), argues for the use of the Unified Modeling Language (UML) on your projects. Years ago, project teams would spend significant time arguing about which modeling notation they should use for their analysis and design efforts — time that would have been better spent doing the actual modeling. With the UML, you no longer have to suffer this problem. Also, a standard modeling language eases communication between developers. There is no longer any need to translate one set of bubbles and lines to another set of bubbles and lines. Fowler also provides good advice about the application of the UML, such as understanding that it is only a modeling language and not a methodology. He also advises that you should use the UML to the level that its readers can handle; there is no use modeling complex constraints if your target audience can't understand them. This is important advice for business modeling efforts because some project stakeholders, particularly your users and senior management, are likely unaware of the nuances of the UML. You need to apply a modeling notation — it had might as well be the standard one.

Your business models likely won't use all of the diagrams of the UML, nor will they apply many of the detailed aspects of the models that you do use.

A key success factor of the Business Modeling workflow, and of all your modeling efforts in general, is that you need to understand how to effectively develop and apply your modeling artifacts. Doug Rosenberg describes a collection of tips and techniques for using the UML on your project in "UML Applied: Tips To Incorporating UML Into Your Project" (*Software Development*, March 1998). The advice presented in this article is applicable to the Requirements workflow, the Analysis & Design workflow, the Infrastructure Management workflow, the Deployment workflow, and of course the Business Modeling workflow. Rosenberg points out that you still need a software process to be successful. Like Fowler, he iterates the fact that the UML is merely a modeling language; it is not a methodology or process. Rosenberg does touch on some methodology issues, advising you to drive your static model from your dynamic model, which in turn, should be driven by your requirements model (use cases, business rules, user interface, ...). To understand how the UML models drive on another, you might want to read "How the UML Models Fit Together" (*Software Development*, March 1998) — an article that I wrote which appears in the Analysis & Design workflow chapter of the Construction phase volume of this series.

A standard modeling notation isn't enough, you need to apply it effectively.

3.3 The Articles

3.3.1 "Essentially Speaking"

by Larry Constantine, November, 1994

Packing everything for a year abroad into suitcases and cartons certainly highlights the difference between "need" and "want," a distinction made sharper by excess baggage charges. In software and applications development, it is also important to get down to essentials — to distinguish the essential heart of what you need to program from the inessential wants and the unnecessary what-ifs.

An essential model is a conceptual tool for focusing the developer's mind on what matters. It represents the core of an application, a problem stripped down to its bare essentials — stripped, that is, of all unnecessary or constraining assumptions.

In software, the notion traces back at least to the origins of structured design. Data flow diagrams were intended as a nonprocedural model of computation, separating the essence of what the computer was to do from how it was to be accomplished through any particular algorithm or procedure. By designing the modular structure to fit this essential definition of the problem, it became possible to build more robust software whose basic form could survive changes in processing details, or so the reasoning went. In practice, designers often turned DFDs into flowcharts with funny figures (and today do the same thing with UML use case models), a corruption encouraged by enhancements that implicitly invited procedural thinking. Essential models finally came into their own with Stephen P. McMenamin and John Palmer's book, *Essential Systems Analysis* (Yourdon Press, 1986), which made them the cornerstone of modern structured analysis.

Essential models capture a technology-free, idealized, and abstract picture of systems grounded in the intentions of users and the fundamental purposes of the system that supports them. The best models, simplified and generalized through repeated refinement, capture the nonphysical spirit of an application, not the physical embodiment in real code on real equipment.

An essential model is based on perfect technology — infinitely fast computers, arbitrarily large displays, keyless input from users — whatever would be needed to most expeditiously realize necessary functions. This technical fantasy is not intended to indulge the imagination but to assure that the model is as independent of current technology as possible. Technology changes much faster than either business practices or people; solutions that are intimately wedded to particular technology are less enduring, less flexible.

Essential Interfaces

User interface design is one promising application. Essential use-case modeling extends Jacobson's object-oriented use cases into user interface design. An essential use case is an abstract scenario for one complete and intrinsically useful interaction with a system as understood from the perspective of the user's intentions or purpose. It is a generalized description of a kind or class of use to which a system may be put, conveying the user actions and system responses in simplified and generalized form. Thus, interfaces are designed to fit user intent, what they want to accomplish, with a minimum of presuppositions about technology, such as the shape of visual widgets or even the devices used for interaction.

For example, consider withdrawing cash from an automatic teller machine. A physical model might take this form: Customer inserts card; system reads magnetic stripe, requests PIN; user types PIN; user keys in selection; system offers menu of accounts; user keys in selection; system requests amount; user keys in amount and presses confirmation button; system spits out cash; user takes the money and runs.

What could be simpler?

Stripped of physical detail and technological assumptions, the essential use case presents a simpler process: customer identifies self, system offers choices, user selects, system gives money, and user takes it. This covers more possibilities for the user interface, including a variety of media for offering choices to the user and for user selection, such as touch-screen or voice response. It highlights that the ATM card and PIN are nonessential; their purpose is to identify the user. Thumbprints, retinal scans, voice recognition, and badge readers might be workable alternatives. The essential model leaves open more possibilities, making it more likely that portions of our design will be reusable as assumptions and conditions change.

This essential use case paves the way to simpler interfaces by highlighting the real heart of the matter to the user: getting the cash, the most common ATM transaction. Most users habitually take the same amount from the same account time after time. The first choice offered by the system could be this user's default, culled from past history: "$250.00 from regular checking, Mr. Chatworth?" or whatever.

The essential model presents an ideal design target. A well-designed user interface requires only as many steps or as much information as is spelled out in the essential use case. The bank customer wants to be able to say, "It's me. The usual. Thanks!" and be off. We specify the ideal case because if we don't model it, we can't design to it. If we don't design to the ideal, we can't see where technology or technical assumptions are limiting us, and we may miss completely the opportunity for alternative approaches.

Re-redesign

Another area of application is business process reengineering — a hot topic in our industry today. When it is not just another euphemism for layoffs, reengineering can be an opportunity to make business processes more efficient and effective. To reengineer a process successfully, you must know what the process is intended to accomplish, what fundamental business or organizational purpose it serves. The single most essential issue in any process or system is the teleological question: Why does this exist? Why should it exist? What is it really for?

Consider exchanging foreign currency at a bank. In some countries this can require lining up two or even three times, with multipart forms, repeated calculation of amounts, and sign-offs by multiple tellers and clerks. From the standpoint of the bank customer, this transaction is ultra-simple: give money in one currency, get an equivalent amount in another. The bank has an interest in assuring that the exchange rates and amounts are correct and in making a small profit on the exchange, but no interest in generating paper or keeping clerks busy — not if the goal is process reengineering.

There is no essential need for a customer or clerk to fill out triplicate forms with names and addresses and signatures and amounts in numerals and text. Exchange rates need not be manually verified if they are derived from one central database; a display facing the customer, as used in many grocery checkouts, serves to validate the incoming amount; a printout of the transaction satisfies the customer's need for a record; the transaction record into the repository gives the bank its audit trail. I've used robot cash machines in Europe that offer precisely this streamlined implementation.

Trance-actions

Unfortunately, many professionals have trouble seeing the essence of their transactions. Not everyone is good at abstract thinking. Thinking in terms of essentials can require setting aside the technical blinders that prevent us from seeing things in a fresh light. Meilir Page-Jones refers to this as "dereferencing," mentally stepping out of the frame of conventional assumptions. It is something that comedians and talented designers are especially good at.

I remember motoring along the Rhein with Meilir as he translated the German signs we passed. One pictograph showed only a black mass sloping down to the left to meet a stack of wavy lines. Canted in mid-air between the representations of the river embankment and the river was an iconic car. "Beware of cars leaping from water on left," Meilir translated.

Dereferencing. Try it.

3.3.2 "The Logic of Business Rules"

by Colleen McClintock, November 1997

Understanding how to collect, specify, and implement business rules will help you build systems and applications that adapt to your company's changing needs.

With our current software development methodologies and engineering practices, business logic tends to get lost in the implementation. In most corporate environments, applications exist on a variety of platforms. They were developed using a variety of methodologies, coded using different programming languages, and they access data managed by different DBMSs. As a result, business logic is buried throughout the systems, leaving applications difficult to maintain and unresponsive to rapidly changing business environments. Removing business rules from the depths of applications and externalizing them so they can be modified easily results in more adaptable applications.

Business Rules Are Everywhere

It seems like everyone's talking about business rules — they're in the middle tier of three-tier architectures, in databases, used as requirements specifications, and so on. Business rules are receiving more press than ever. They're mentioned regularly in technical periodicals, and publications such as *Business Rules ALERT!* and *Database Newsletter* are dedicated to them. Vendors of many unrelated products are claiming support for business rules — some are even calling their products "business rule tools."

When people discuss business rules, they may not always be referring to the same concept. Generally, they're referring to an implementation of business policy in computer systems or a modeling construct used to represent business logic. To define a business rule in this article's context, I'll quote my colleague, Margaret Thorpe. In her article titled "Understanding Business Rules" (*Business Rules ALERT!*, June/July 1997), she writes, "A business rule is an atomic, explicit and well-formed expression that describes or constrains the business principles, guidelines and operations of a company using vocabulary and syntax that can be easily used and understood by the persons within the company who are responsible for defining and carrying out the business."

Organizations have thousands of such business rules that define and direct how they manufacture, sell, buy, and otherwise transact business. Often, these rules are defined in regulatory and policy manuals. An example of a business rule might be: "For any given life insurance policy, the beneficiary cannot be the same as the policy holder."

Collecting and Managing Business Rules

Systems and data analysts have informally captured business rules for some time. Data analysts have captured them in the form of constraints including optionality, cardinality, and referential integrity. More complex constraints restricting more than one entity, attribute, or relationship could not be easily represented in data models and usually were captured in text or pseudo-code, then passed on to the application programmers for implementation.

When the representation and process for collecting and implementing business rules are formalized, the rules become a valuable tool for specifying requirements and communicating with business people. This perceived value has led many companies to begin actively collecting them. You can extract business rules either from legacy applications, or collect and define them for new applications using business policy documentation and business experts as sources. Tools such as ReGenisys's extract:R and ruleFind:R (http://www.regenisys.com) can assist you in finding business rules in legacy applications. However, all related business rules may not be automated in legacy systems, and even those that are may not be complete and correct. Regardless of whether business rules are being collected for new applications or extracted from existing ones, business users should be involved in the process.

To formalize the representation of business rules, you can define a formal grammar for a specification language. Although business rules can be captured and represented using structured English, using a business rule specification language ensures the rules are unambiguous and specified using consistent syntax. In short, the business rule specification language is computable.

The idea behind a specification language is that specification and implementation should be kept separate. The specification should state what the software is supposed to do rather than how it will do it. Although no business rule specification languages are widely used, several companies have defined their own specification language to support business rules and several products use their own proprietary languages.

Once they're collected and specified formally, you can store business rules in a repository. A repository provides the flexibility to store, retrieve, validate, represent, and manipulate the organization's relevant meta-data, including business rules. With a repository, it's possible to track where business rules are represented in the business models, as well as where they are implemented within databases and applications. With a well-designed repository, you can easily perform impact analysis and reduce the turn-around time for business policy changes.

The repository can simply be a database used to store business rules meta-data, or a commercial repository such as Rochade Repository (http://www.rochade.com), Microsoft

Repository (http://www.microsoft.com), or the Platinum Open Enterprise Repository (http://www.platinum.com). If you use a commercial repository, it should be extensible so you can accommodate a business rules meta-model. The meta-model is based on the business rule specification language and should store the business rules in a format that supports the grammar. In addition to the business rules themselves, a repository should store relevant business rule meta-data such as rule owner, status, and source.

To provide users with access to business rules, you can develop a repository user interface, as shown in Figure 3.2. This tool is a key component for automating the management of business rules. It lets users define, browse, and query business rules.

Figure 3.2   Components of a business rule server environment.

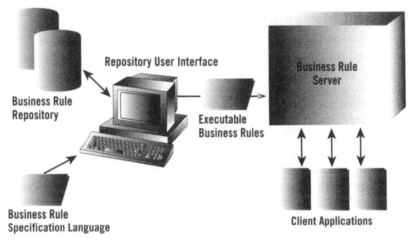

Incorporating Business Rules into Your Modeling Approach

Although current methodologies provide varying levels of support for business rules, no comprehensive business rule modeling approach exists today. Historically, function-oriented, data-oriented, and object-oriented methods have lacked a convenient way to specify business rules. This is improving with more recent modeling approaches such as the Unified Modeling Language. Part of the confusion surrounding the term "business rule" stems from the lack of a formal methodology for collecting and representing them.

Nonetheless, collecting and specifying business rules can complement any systems development methodology. Business rules are central to all phases of the system development life cycle. For example, in object-oriented analysis and design, you can incorporate business rules from the requirements phase through the implementation phase. Use cases are a mechanism for capturing system requirements during the analysis phase. You can capture business rules using structured English at this early phase. As class diagrams begin to emerge, domain classes, attributes, methods (actions), and relationships are identified and represented. At this stage, you can specify business rules more formally by using the constructs of the class diagram. Objects, attributes, values, and actions referenced in rules should originate from class diagrams. As analysis progresses and class diagrams mature, you can use business rules to specify more complex integrity constraints.

Class diagrams, although central to object-oriented analysis and design, are not the sole representation of the system. The dynamic and functional views are modeled in sequence diagrams, state diagrams, activity diagrams, collaboration diagrams, and so forth. You can also capture business rules that reference the system's actions and events. As you collect and specify business rules, validate and refine them with business experts. This process should be iterative, along with the refinement of the various object-oriented analysis and design models. After they're validated, business rules can be used in the system design and implementation.

Approaches to Implementation

Historically, most business rules have been embedded in the logic of applications programs, as this is where the procedures for checking and enforcing them exist. Today, the three main approaches for implementing business rules use databases, objects, or a business rule server. Each of these approaches is valid for different types of business rules and appropriate for different technical architectures. It's unlikely that a single implementation approach for all business rules will emerge in the foreseeable future.

Listing 3.1   Simplified business rule server interface using CORBA IDL.

```
//Declarations of data types


    typedef any FieldValue;
    typedef string AttributeName;


//AttributeStruct contains the attribute name and its value


    struct AttributeStruct  {
        AttributeName    nameofAttribute;
        FieldValue       attributeValue;
    };
    typedef sequence<AttributeStruct> Attributes;
    typedef string ObjectName;


//InstanceStruct contains the attributes and values for each object instance


    struct InstanceStruct     {
        ObjectName           nameofObject;
        Attributes           objectAttributes;
    };
    typedef sequence<InstanceStruct> Instances;


//ObjectStruct contains the instances for each object, allowing more
//than one instance of an object to be passed to the server
```

```
    struct ObjectStruct  {
        ObjectName              nameofObject;
        Instances               objectInstances;
    };

//BusinessObjects is a sequence of objects

    typedef sequence<ObjectStruct> BusinessObjects;

    typedef unsigned short RuleID;
    typedef unsigned short RuleVersion;
    typedef string RuleText;

//ResultStruct contains the result of rule evaluation

    struct ResultStruct {
        RuleID                  businessRuleID;
        RuleVersion             businessRuleVersion;
        RuleText                businessRuleText;
    };
    typedef sequence<ResultStruct> BusinessRuleResults;

//Business Rule Server Interface Definition

    interface BusinessRuleServer  {
    BusinessRuleResults evaluateRules (in BusinessObjects busObjects);
    };
```

Database-Centric Implementation

Integrity constraints are frequently implemented in relational databases using triggers and stored procedures. However, not all business rules lend themselves to this kind of implementation. Some may not be applicable across all applications sharing a database, but instead may be applied and enforced differently within various applications. It is the business context, implemented within applications and objects, that usually determines how and what business rules apply within a particular transaction.

Newer, model-based business rule development tools are an alternative to manually implementing business rules in a database. Two such tools are VisionBuilder from VisionSoftware (http://www.vision.com) and USoft Developer from USoft (http://www.usoft.com). Both tools allow automatic generation of executable application components from a declarative business rule specification language. VisionBuilder enforces business rules in the presentation layer in Visual Basic or Java, and in the data layer as Oracle or SQL Server triggers and stored procedures. USoft Developer uses a "rules engine" to enforce business rules. Through the USoft ODBC Ruler, the rule engine can control access to any DBMS, ensuring all relevant business rules are enforced.

Object-Centric Implementation

There are several approaches to implementing business rules in an object-oriented environment. However, there are fundamental differences between rule-based and object-oriented programming. Theoretically, rules violate the principle of encapsulation, since they must reason across objects. In addition, the object-oriented paradigm models and implements all aspects of an application procedurally, whereas most business rules are more naturally represented declaratively. These conflicts introduce complexity into the representation and implementation of business rules in an object-oriented environment.

The most obvious and probably most common approach is to encapsulate business rules in methods of entity objects where you must code them procedurally. This works sufficiently well when a single method in a single class can implement the business rule. However, when the business rule must be implemented in several methods — especially across multiple classes — the model, and subsequently the code, becomes difficult to understand and maintain.

Another object-oriented approach is to represent each business rule as an independent object, since each has properties of its own. This approach can lead to performance and implementation problems since, due to encapsulation, a lot of messaging may be required between objects to obtain the data values needed to evaluate the rules.

A promising approach is to embed rules in intelligent objects, thus encapsulating them in the objects that use them. Declarative rules, which are processed in a data-driven manner, are integrated into an object-oriented environment. Researchers and developers at AT&T Bell Laboratories have created an extension to the C++ language, called R++, that supports the encapsulation of rules in C++ objects. You can find vendor support in the ILOG Rules (http://www.ilog.com) or Rete++ (http://www.haley.com) products, which let you integrate data-driven rules with C++ or Java objects.

Business Rule Server Implementation

In a business rule server environment, client applications request services of the business rule server, as shown in Figure 3.2. The business rule server contains business rules in executable form. Client applications pass data to the server and the results of the business rule evaluation are returned. A number of other components exist, which when used in conjunction with a business rule server, can greatly enhance this approach. These include the repository, where business rules are managed and stored; the repository user interface, which gives users access to the rules; and the business rule specification language, from which you can generate executable business rules.

A business rule server can be a flexible and scalable approach to implementing business rules. A rule server is particularly well-suited when business rules are based on business policy that is subject to frequent change, are complex, or are shared by several applications.

Business rules are often subject to frequent changes. They may represent policy imposed either internally in the form of directives to achieve strategic goals, or externally by legal and regulatory agencies designed to realize public or economic strategies. The responsiveness of the systems in implementing these changes frequently dictates the competitive position of the organization. Implementing the business rules in a single place, rather than throughout the applications, greatly simplifies the modifications required to implement policy changes. Also, a business rule server simplifies the task of externalizing and managing business rules at the specification level rather than the code level.

Complex business rules can be difficult to represent and maintain using procedural code. To ensure correct enforcement, you may need to duplicate these business rules in a variety of places throughout the code. This can lead to omission and logic errors as the application is extended. A business rule server implements a business rule as a declarative construct that has a one-to-one correspondence with the business rule represented in the specification language or is generated from that business rule.

Frequently, business rules are implemented in multiple applications. This is often the case when certain rules apply across multiple business processes. For example, in the secondary mortgage market domain, a large number of rules pertain to the eligibility of a loan for purchase by a secondary mortgage institution. These rules apply to different business processes including the origination, underwriting, and purchasing of loans. Typically, different applications perform the eligibility evaluation for each business process. Thus, each must contain the appropriate business rules. A business rule server lets applications share these rules. Rather than duplicating the rules in each application, applications request the services of the business rule server.

Using a Commercial Rule Engine

At the heart of the business rule server is the rules engine. Its responsibility is to evaluate the executable business rules. The rules engine is a commercial product, commonly known as an expert system shell, that supports rule-based programming. Most rules engines are designed for stand-alone development of rule-based applications. They offer features such as proprietary procedural programming languages, GUI builders, and database integration. For the purpose of implementing a business rule server, however, two key components are of interest: the rule language and the inference engine.

Rule engines represent declarative rules in the form of IF..THEN rules. The language used by the rules engine supports one-to-one correspondence with a business rule represented in the specification language, thereby simplifying coding or generating executable business rules. When business rules are coded procedurally, they often result in complex switch and nested if statements where the evaluation order is important to achieving the desired result. A rules engine separates knowledge and control by representing the knowledge of the problem domain in declarative rules that are evaluated by the inference engine. The inference engine contains the procedural knowledge about how and when to apply the rules.

Most of the commercial rule engines have inference engines that are data-driven, meaning they evaluate rules in response to changes in data values. This approach naturally supports the processing required in a business rule server environment in which client applications pass data to the business rule server, and the server returns the results of the business rule evaluation. Also, the inference engines of these products are highly optimized for efficient rule processing. They implement special pattern-matching algorithms for efficiently evaluating the conditions of rules in relation to changes in data. Without such an algorithm, each change to the data that the rules act upon would cause reevaluation of the entire set of rule conditions.

Distributed Processing Environment for a Business Rule Server

Most commercial rule engines do not provide a built-in server layer to facilitate the deployment of rule-based applications as servers. Many are oriented toward two-tier client/server applications and do not promote sharing business rules across applications. Therefore, developing a server layer is usually required to implement a business rule server.

For maximum flexibility and reuse, a business rule server should be developed as a stateless server. Once it processes a request, it reverts to its original state. In the case of a business rule server, once the rules have been evaluated and the results have been returned to the client, the transaction is complete from the server's standpoint. The server has no knowledge of previous client requests. Stateless servers are better suited for most distributed processing environments; CORBA, for example, does not have a mechanism for maintaining server state.

A business rule server works best in a service-based distributed processing environment. Ideally, the server is not aware of how the client application made the request, what language or platform the client application is communicating from, or where the client is located. Products such as object request brokers and message-oriented middleware provide an optimal communications infrastructure. For a good overview of distributed computing issues and middleware alternatives, see Tom Laffey's article titled "Making the Message Clear" (July 1997).

You can also develop a communications infrastructure using sockets or remote procedure calls. In an early business rules project I was involved in, in 1994, we used Sun ONC RPC to develop the communications layer of the business rule server. This is an acceptable solution if your organization currently lacks a distributed processing architecture. However, when using sockets or remote procedure call mechanisms, take into consideration the future architecture standards your organization may be evolving toward.

Support for Business Rules in the UML

The Unified Modeling Language (UML), represents the unified methods of Booch, Rumbaugh (OMT), and Jacobson. The UML is a modeling language (consisting mostly of graphical notations) and not a methodology. As a modeling notation used to express designs, the UML can be used by any methodology.

Unlike formal methods, in which designs and specifications are typically represented using predicate calculus, set theory, and relational calculus, object-oriented methods have little rigor. The UML, which is based on a meta-model, injects a new level of formality into object-oriented methods. The idea behind the meta-model is to more precisely define what is meant by such concepts as association, generalization, and multiplicity. The meta-model is specified using the UML itself.

Support for business rules in the UML comes mainly in the form of constraints. The UML considers constraints to be "a semantic relationship among model elements that specifies the conditions and propositions that must be maintained as true (otherwise the system described by the model is invalid)." The class diagram itself contains many constraints including the basic constructs of association, attribute, and generalization. Constraints are put inside braces ({ }). In the UML, certain constraints such as "or" are predefined. The UML has 14 such predefined constraints.

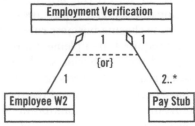

This UML "or" constraint specifies that Employment Verification must consist of either an Employee W2 form or two or more pay stubs.

A constraint on a single graphical symbol, such as a class or association, is placed near the symbol. A constraint involving two graphical symbols is shown as a dashed line or arrow labeled by the constraint in braces. When three or more graphical symbols are involved, the constraint is placed in a note symbol and attached to the symbols using a dashed line.

UML 1.0 does not prescribe the language used to specify constraints. It could be informal English, structured English, predicate calculus or a derivative, or even a segment of program code. However, UML 1.1 and UML 1.3 both contain an expression language for specifying constraints, Object Constraint Language (OCL). Hopefully, we can expect support for OCL in future versions of the Rational Rose modeling tool and other UML-based modeling tools.

More information and the most recent updates on the UML and OCL are available at `http://www.rational.com/uml`.

Interface to the Business Rule Server

The API to the business rule server should be adaptable. Through the API, client applications pass the data upon which the business rules are evaluated and the results of rule evaluation are returned. Each client application will most likely need to pass different data to the server. The API must be flexible enough to accommodate the data passed by any client so that regardless of the data being passed, all clients use the same API. This will enable new client applications to use the business rule server without changes to the API. In addition, the API should be independent of business rule changes. Modifications to business policy that result in the addition, deletion, or modification of business rules should have no impact on the server interface.

You can accomplish these objectives by using a self-describing data structure in the API, as shown in Listing 3.1. The API is described using CORBA IDL because it is the open industry standard for defining software interfaces. Similar business rule server APIs have been implemented in distributed processing environments that are not object-oriented.

Using the API, clients send objects to the business rule server. The server must interpret the objects in the data structure and create instances of those that are visible to the rules engine. Commercial rule engines that can inference directly across C++ or Java classes simplify this task. Other rule engines require the data be mapped, using a C or C++ API, into a proprietary object representation internal to the rule engine environment.

For performance reasons, the business rule server API lets clients pass objects across a CORBA interface rather than calling them remotely. In a purely C++ CORBA environment, where all clients are written in C++, an alternative to this type of API would allow the client to transparently send C++ objects to the server using a product such as RogueWave's ORB-Streams.h++ (http://www. roguewave.com).

Once the objects are introduced to the rules engine, the business rules are evaluated. The results of business rule evaluation are returned to the client application in the results data structure shown in the API. Enforcement of the business rules, in the form of actions, remains the responsibility of the client application. This is appropriate, since the action is frequently dictated by the context or business function of which the client is aware.

Business Rules Empower the Business

Central to our information systems are the policies and procedures that are fundamental to our business. As these policies and procedures evolve, new systems are built and existing systems are modified. Recognizing the fundamental importance of business rules and understanding the options for implementing them helps us build more intelligent systems that empower rather than hinder the business.

3.3.3 "Why Use the UML?"

by Martin Fowler, March 1998

What's the second biggest news in the object-oriented development world at the moment? It's the Unified Modeling Language (UML), which will replace all of those analysis and design methods (Booch, Coad, Jacobson, Odell, Rumbaugh, Wirfs-Brock, and so forth) with a single new notation endorsed by the Object Management Group (OMG). This will finally eliminate a tedious range of arguments about which method software developers will use. We will enter a new period of harmony, brotherly love, and increased productivity (and have a good snicker at those old structured methods which, despite their greater maturity, never achieved such a standard).

If you were a keen user of one of these B.S. (before standardization) methods, you'll bite the bullet and switch over to the UML, muttering darkly about how it missed feature x from your favorite method and how you don't like features y and z cluttering it all up. But have you really stopped to think about why a modeling notation such as the UML is useful?

Ask a methodologist (in our industry that's someone who invents a methodology) and you'll get a stern lecture on software quality. Methodologists will talk about how our industry suffers from the software crises, problems of poor software quality, the importance of good design, and so on. This is all well and good (although I think the software industry has done pretty well over the last decade), but how exactly does the UML help? Ask a CASE tool vendor and you'll get a lecture on improved quality, automatic documentation, and the productivity value of code generation. But we all know what CASE tool vendors are after.

If you're a developer who is suspicious of all this methods stuff as yet another management fad, and you shudder at the thought of all the useless paper that gets generated, then the UML is just another notation to pretend to care about. At least there's only one now. Still, you know that the next time you have to modify a system, these UML diagrams that are supposed to be so helpful will be several generations behind the code you have to fix.

I got into methods early in my software career. Since I have an engineering background, they seemed to be a natural field of interest for me. Most branches of engineering have drawings that describe how things should be built, and you take great care that things are done according to the drawing. I saw methodology diagrams in the same way. In time, I learned the lie of that analogy, but I still find methods useful. This despite the fact that I have a lot of sympathy for developers who dislike methods.

To describe why I find the UML useful, I first must start with a quick reminder on the scope of the UML: it's a modeling language, not a method (or methodology). The UML defines a number of diagrams and the meaning of those diagrams. A method goes further and describes the steps in which you develop the software, what diagrams are produced in what order, who does what tasks, and so on. The idea behind the UML is that it is method-independent. Over the next year or two, we will see various people come up with methods that use the UML notation. But you don't have to use a method to make use of the UML. In this article, I'm not going to assume any fixed method of development.

So if we strip away all the method trappings, what are we left with besides a handful of types of diagrams that you can draw? The question, "What use is the UML?" becomes "What use are these diagrams?"

Figure 3.3   A UML container of Java's AWT container classes.

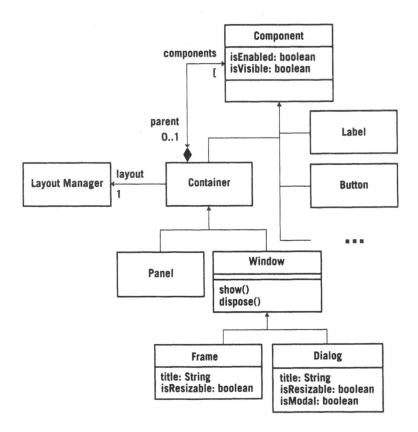

The answer boils down to one word: communication. This is an important word. The main reason software is deceptively hard to develop is communication. We know how easy things would be if we could only sneak off for a weekend and code away. The difficulty arises because we must communicate with multiple developers. The UML is important because it can help software developers communicate. We have to use it in a way that helps communication and does not hinder it.

Figure 3.3 is an example of how the UML can help communication. It's a diagram that begins to describe the Container class in Java's Abstract Windowing Toolkit (AWT). By reading the diagram, I can understand a lot of things. I can see that Container is a subtype of Component. Components can be made visible or active, and other kinds of components include labels, buttons, and more. I can ask a component for the container in which it is contained, but not all components need a container. Containers include components (which may be other containers) and also have a layout manager. Container, like Component, is an abstract class. Its subtypes include panels and windows. Windows can show and dispose themselves. Window also has subclasses frame and dialog. Both frames and dialogs have titles and can be set to resize or not. Although both subclasses of window do this, this behavior is not part of window itself. Dialogs can be marked as modal, but frames cannot.

You might prefer the diagram or you might prefer what I said in the previous paragraph. This depends on whether you are familiar with the UML and whether you prefer a visual or narrative statement. For this situation, I prefer a visual, but there are those who prefer the text, even if they understand the diagrams. You could give them the text or (perhaps better) a selection of code (as shown in Listing 3.2). Which would you prefer? Which do your colleagues prefer? These questions are the crux of the role of the UML and similar languages. I've found that some prefer text, others prefer the diagram, some prefer the diagram for some things and the text for others. In the end, the diagram is only worthwhile if it enhances communication.

In addition to what the diagram shows, you should also note what the diagram does not show. All of the classes mentioned have much larger interfaces than either Figure 3.3 or Listing 3.2 indicate. I've also failed to mention that Layout Manager is an interface, or that Component implements a number of interfaces. Many would thus criticize Figure 3.3 for being incomplete. Incomplete it is, but is that a failing? In the diagram, I made a decision about which features of the depicted classes to show and deliberately did not show others. The fact that I show only some of these features highlights the features that I've shown.

Selecting information to give it emphasis is an important part of communication. In any group of classes, there are some aspects that are more important to understand to gain an initial comprehension of the classes. If you show everything, you fail to make that distinction, and your reader has no idea what is important to understand first and what is detail to be absorbed later. When I use class diagrams, I use them for this initial comprehension, to understand the key aspects of the classes about which I'm talking. I know that the reader can always go to the javadoc files to get a complete statement of the interface.

I encourage you to use class diagrams in this selective manner. Not only does this improve the communication value of the diagrams, it also makes them easier to keep up to date. You don't have to change the diagram for every little change in the class. Since one of the biggest problems with these kinds of diagrams is the difficulty of keeping them up to date, this is an important advantage.

Listing 3.2   Java interfaces for Figure 3.3.

```java
public abstract class Component {
  public Container getParent();
  public boolean isEnabled();
  public boolean isVisible();
  public void setEnabled (boolean b);
  public void setVisible (boolean b);
  …
}


public class Button extends Component {…}
public class Label extends Component {…}
// other component subclasses
public abstract class Container extends
  Component {
  public Component add (Component comp);
  public void remove (Component comp);
  public Component[] getComponents();
  public LayoutManager get Layout();
  public void setLayout(LayoutManager mgr);
  …
}

public class Panel extends Container {…}
public class Window extends Container {
  public void show();
  public void dispose();
  …
}

public class Frame extends Window {
  public String getTitle();
  public void setTitle(String title);
  public boolean isResizable();
  public void setResizable (boolean b);
  …
}
```

```
public class Dialog extends Window {
  public String getTitle();
  public void setTitle(String title);
  public boolean isResizable();
  public void setResizable (boolean b);
  public boolean isModal();
  public void setModal();
  ⋮
}
```

As well as encouraging selectivity, I also encourage an emphasis on interface rather than implementation. I've shown an attribute of isEnabled on Component. This doesn't say that the Component class has a field called isEnabled (I actually didn't look, because it doesn't matter). It says that from the outside, you can assume the class has such an attribute, which you would access through the appropriate operations. Ideally, there will be a naming convention for these operations (in the Java library these days, this is isBooleanAttribute and setBooleanAttribute). I don't show the operations on the class because I find the attribute notation communicates the intention of the code better. This attitude extends to associations too. I have no idea what data structure exists between Container and Component, the operations suggest the association.

Many people draw class diagrams from an implementation perspective: the attributes and associations to reflect the data structure. This is valuable if the data structure is what you wish to communicate. Often, however, the interface is what's most important. You should decide what it is that you wish to communicate and follow that perspective.

I find diagrams are also useful when discussing a design and how you might change it. If you have a group of designers working on a design, try sketching the designs on a white-board. Draw a few alternatives. I find it's a valuable way to visualize what we're talking about. (CRC cards are another good technique for this.)

A particularly important variant of this technique is when I am working with a domain expert to try to understand the domain for which we are building a system. In this situation, I use minimal notation and concentrate on the concepts in the domain expert's head rather than any particular software situation. I've found it easy to teach this conceptual modeling style to people without a software background. Then, using the diagram, we can jointly develop a well-defined vocabulary to talk about the domain and can come up with abstractions that are useful both in discussions and in the final software. This was a great boon to me when I worked with complex domains such as healthcare and financial trading.

Standardization is valuable here because it enhances communication. It's difficult to communicate with people when they use a variety of diagramming styles. By having a single standard, we can be sure that if people know any diagraming styles, they will know this one. But don't let this go too far. The UML contains a lot of notation, and there's no rule that says you must use all of it. Try to use a fairly minimal part of the notation, and don't use the advanced concepts unless they're really necessary. Although you should stick to the standard as much as you can, I'll admit I'm not afraid of bending the notation if need be. I don't do it often because each bend has to be explained and is going to be unfamiliar to the reader — but if it enhances communication, I do it.

So if you're new to the UML, try it out as you need to communicate ideas. Experiment to see what works and what doesn't. Learn the notation by using it, and learn it gradually. If you're more experienced with modeling notations, you should have no difficulty adapting to the UML — but beware of taking it too far. Remember its primary purpose. Be aware of how well your diagrams are communicating. Don't leap into more complex use of the UML than your readers can handle. And remember to be selective and to highlight important information.

3.3.4 "UML Applied: Tips To Incorporating UML Into Your Project"

by Doug Rosenberg, March 1998

In the last quarter of 1997, I taught object-oriented analysis and design with UML to a wide range of developers working on various projects. The projects spanned application domains including retail shelf space management, pharmaceutical clinical trials, cellular telephony, and more. Implementation languages included Visual Basic, C++, Java, DB2, and others. This article discusses several aspects of how well the UML notation, along with my company's use case-driven Unified Object Modeling process (also based on the methods of Booch, Jacobson, and Rumbaugh), met the demands of a diverse set of projects. It provides practical guidance for tailoring your use of UML for various projects.

The UML notation is big (maybe too big) and is flexible enough to accommodate the needs of a very wide range of projects. To succeed with UML, you must streamline how you use it. Different projects will need different pieces. Which specific elements of the UML notation you need for your project will depend on its nature (client/server with a mainframe RDBMS vs. real-time embedded, for example) and on the implementation language you will be using. Some detailed C++ design constructs are not needed if you're building in Java or Smalltalk, and you may want to avoid too much use of generalization or inheritance if you're building in Visual Basic. Sometimes, the sheer volume of modeling constructs can be overwhelming, especially to those students who are new to object-oriented analysis and design. However, the good news is that you can model almost anything using UML. There are plenty of modeling constructs to go around.

You Still Need a Process

UML itself is only a notation; to build effective models, you still need a modeling process. There are many modeling approaches you can use with UML; the modeling language itself does not prescribe the process. My company has been successful teaching a slightly modified version of the Objectory process[1] (derived from Ivar Jacobson's *Object-Oriented Software Engineering: A Use Case Driven Approach*, Addison-Wesley, 1992) that we developed as a synthesis of Booch/Rumbaugh/Jacobson several years prior to the advent of UML. Our approach places slightly more emphasis on identifying high-level static models (domain models) up front, in parallel with use cases. We then refine the static model iteratively and incrementally refine as we explore the use cases and dynamic models.

---

1. Which has since evolved into the Unified Process.

Whether you prefer using Objectory, Object Modeling Technique, our ICONIX Unified Object Modeling process, or some other approach, it's important to understand that UML is not a process, and that it is critically important to get your project team on the same page process-wise before undertaking a modeling effort. Once again, the size and bulk of UML (and the overwhelming weight of documentation on notation as opposed to process) can make it easy to slip into "analysis paralysis." Focusing the team on an understandable process that is supported by the UML notation can generally get the modeling effort underway.

Legacy Methods Are Important

Many of my students ask whether developing an understanding of Booch, Jacobson, and Rumbaugh's legacy methods is still important. My answer is an emphatic yes. Just as knowing the symbols used for electronic circuit design doesn't eliminate the need to know circuit theory, understanding the notational aspects of UML doesn't eliminate the need to understand object modeling theory. Since UML represents the synthesis of the works of Jacobson, Booch, and Rumbaugh, the original legacy methods are a rich source of this information.

Keep It Simple

Getting a project team to make effective use of UML is tricky. Team members will have varied experience with object-oriented analysis and design and UML. A training workshop is a good way to begin. The workshop needs to be tailored to the specific needs of each project, taking the backgrounds of the various team members into careful consideration. The most critical tailoring decisions will ultimately involve judicious choices of what gets left out of the course agenda, as well as the instructor's ability to adjust on-the-fly during the workshop.

One of the most important things to remember when learning UML is that you don't need to use every construct just because it's there. Keep the process as simple as possible. Streamlining a project's documentation set and modeling approach does wonders for generating forward momentum.

Modeling with UML is similar to sitting down to an huge plate of food — sometimes, the thought that you can't possibly eat everything on the plate just kills your appetite before you get started. A similar phenomenon can occur with UML modeling. The thought of having to produce a complete set of sequence, collaboration, state, deployment, use case, and class diagrams that comprehensively covers each and every use case of the system with a fully detailed dynamic model can intimidate a team right out of object-oriented analysis and design.

The same thought process holds true for determining which constructs are employed within a given modeling technique. For example: is it really necessary to employ both USES and EXTENDS on a use case diagram (in UML 1.2+, now called INCLUDE and EXTEND respectively), or can we live with just USES? My experience has been that the more streamlining that gets done, the better the chances of the modeling effort being carried through.

Write the User Manual Before You Design the Classes

One of the old saws of programming is to write the user manual before you write the code. This was good advice when I learned it, and it's still good advice today. In the days of structured methods, and in the early days of object-oriented methods, it was seldom followed. In his use case-driven modeling approach, Jacobson codified this maxim into a series of steps that work for object orientation and can be described using UML. Each step builds upon the previous step in a traceable manner, so that ultimately, management can enforce this approach as a design paradigm and verify that it has been followed at each step in the analysis and design process.

The key to understanding the essence of Objectory and use case-driven object modeling at the fundamental level is simple: write the user manual before you design the classes. Keeping this idea in the front of your mind will help guide you as you travel through the maze of UML static and dynamic model notations. Each use case represents a portion of the user manual, and should be written that way if the goal of your use case analysis is to produce an object model.

Organize Use Cases into Packages

As you begin to write the user manual for your system one use case at a time, you will immediately run into the need for a high-level organization of use cases. UML lets you organize your use cases into packages. These are represented as tabbed-folder icons. Each package should consist of at least one use case diagram, which will serve as a context diagram under which you can link all the use case descriptions along with the design-level dynamic model views for each use case. Some projects start with a top-level breakdown of one package per top-level menu. While this breakdown does not always represent the final breakdown, it's sometimes a helpful starting place.

Use the Objectory Stereotypes

Since we're driving the entire design process from the use cases, it makes sense to focus strongly on describing them in the "right" way. While it's becoming increasingly popular to employ use cases as an alternative to requirements specifications and for business process modeling, and while these styles of use case modeling have somewhat different guidelines, most projects I've run across still view use cases as a way to get to an object model.

Jacobson's original OOSE/Objectory process included a phase called Robustness Analysis, wherein use case descriptions were analyzed to determine a rough first cut at a collaborating set of objects. While doing this analysis, Jacobson proposed classifying the objects identified into Interface Objects, Control Objects, and Entity Objects.

This small, quick modeling step produces an amazing set of benefits. It helps people write use cases correctly, identify client/server and model-view-controller design information early, and perhaps most important, identify reusable objects by enabling a quick breadth-first pass across all the use cases. It also fills the void between requirements analysis and detailed design.

Unfortunately, for some reason, the notation for Robustness Analysis (three easy-to-draw symbols), only partially survived the transition into UML. The notation still exists, but has been banished to a separate document called the Objectory Process Specific Extensions, and tool support is often lacking. I teach Robustness Analysis as an integral part of describing use cases (the diagram becomes a sanity check for the text), and have found that students readily adapt to this object shorthand.

Important Questions

You can reduce the entire domain of object-oriented analysis and design to two simple questions: First, what are the objects in the system? Second, how is the required system behavior distributed among these objects?

While this is somewhat of an over-simplification of a subject that has been the topic of thousands of pages of methodology textbooks, it fundamentally isn't too far off the mark. If you've identified the right set of objects to build, and if you've done a good job of allocating the desired system behavior to the most appropriate set of classes, your project should be in good shape. The tricky part is really the innocent-sounding phrase "done a good job of allocating behavior"; this is where experienced object-oriented designers earn their living.

Drive the Static Model from the Dynamic Models

No matter which process you decide to use with UML, it's good practice to drive the design of the static model (class diagrams) from the dynamic models, starting from use cases at the high level, and particularly leveraging the UML Sequence Diagram to allocate behavior among your classes. This philosophy, which has its roots in Jacobson's OOSE/Objectory process, was first explained to me around 1993 by a friendly Objectory consultant. As I've continued to teach it as a design style over the last four years, I've grown increasingly convinced of its wisdom and (thus far) universal applicability.

The essence of the idea is this: we can start out and get a rough approximation of the core set of objects in a system by following an Object Modeling Technique-like strategy of identifying "real-world" or "problem domain" objects using techniques such as looking for nouns in the problem statement. Sometimes, we can make intelligent guesses as to when a particular class might be an appropriate container for a specific operation; however, often in the process of object-oriented design, we find that the original guesses that we made when considering the static models in the absence of exploring the use cases were naive.

Based on my experience, the reality of object-oriented analysis and design is that the only really good way to approach the (difficult) problem of allocating behavior among a set of classes is to explore the use case at a detailed (message passing/method invocation) level. Whether formally or informally, most senior object-oriented designers I've met arrive at their designs this way. When the approach is informal (not visible), a cloud of mystery sometimes surrounds how a designer arrived at a specific solution from a given set of domain objects and use cases. Often, this cloud of mystery is deepened by oblique explanations using a litany of jargon such as "multiple polymorphic encapsulated interfaces." This tends to limit the amount of useful design review that can be accomplished by team members and leaves the intrepid programmer free to implement as he or she sees fit.

In my experience, however, this design approach is best accomplished using a sequence diagram in UML with the original (requirement-level, user manual view) text of the parent use case shown along the left margin, and the actual detailed dynamic behavior, showing each method invocation and the message that invokes it, in the body of the diagram. Showing the detailed design and the requirement-level textual use case description on the same page provides a quick "eyeball" requirements trace, which verifies that, for at least this use case, your design matches the requirements. Simply repeat this for all the use cases of your system, and you have a traceable design.

While drawing the sequence diagrams, you'll identify specific operations and assign them to specific objects within your design. You're actually building the static class model, even though you're drawing dynamic model (sequence) diagrams at the same time. The sequence diagram is the vehicle that teaches us how much we don't know when we explore the object model in the abstract.

Defer Assigning Operations to Classes

Don't worry too much about specifying which operations go in which classes during the analysis phase of your project. You'll probably get the mapping wrong anyway, except in the most obvious cases (and sometimes even then). Experience teaches that these behavior allocation decisions are best made very carefully, one at a time, as the dynamic models are explored.

Keeping this separation in mind (Analysis: what are the objects? Design: how is the behavior allocated?) helps project teams define the boundary between analysis and design. Our original Unified Object Modeling process approach used the Object Modeling Technique notation at the analysis level, and the Booch notation for design. The Object Modeling Technique was applied during analysis using the Booch method for detailed, design-level models. With UML, these notations have been merged into a single, Object Modeling Technique-flavored class diagram. As the line between analysis and design notations has blurred, project teams often experience difficulty in understanding where the analysis and design boundary is.

Even though I teach an iterative and incremental process, a requirements review and a design review must be included at some logical point. If I'm reviewing an analysis model, I'm not particularly concerned whether or not the classes have operations showing (in most cases, I'm just as happy if they don't). I'm looking for a good set of domain classes and a comprehensive use case model. For a design review, however, all operations should be accounted for, and there must be visual traceability between the requirements and the design on the sequence diagrams. The designer must also be able to defend his or her decisions as to why any given operation was placed in a particular class.

Simply Successful

The most important things to keep in mind as you apply UML to your project are to keep it simple, write the user manual first (one use case at a time), and get your project team on the same page regarding process. Remember, UML specifies a notation, not a process. The most successful projects I've seen are adopting use case-driven, iterative, and incremental approaches. If you tailor your process to the individual parameters of your project and to the skill set of your team, your UML project will be marked for success.

Chapter 4

Best Practices for the Requirements Workflow

The goal of the Requirements workflow is to engineer the requirements for your system to the "80% point" — the point at which you generally understand what the system is supposed to do but you still have some details that you need to research and understand. The purpose of requirements engineering is to determine what your customer actually wants, not what you *think* they want. The fact that you may have more than one group of customers for your system often complicates your requirements engineering efforts. For example, perhaps you have line workers that will use your system to do their daily jobs, their managers use portions of your system to facilitate their employee's efforts, and the marketing department uses sales information generated by your system. Another complication that you need to overcome is the fact that you often don't have access to your users/customers — which experience often proves to be a leading cause of failure for many projects.

During the Elaboration phase, your goal is to evolve your requirements model to the 80% completion point.

There are many ways to engineer the requirements for your system, as you can see depicted by the solution for the *Define and Validate Initial Requirements* process pattern (Ambler, 1998b) of Figure 4.1 (the techniques of which are described in section 4.4.1 "Engineering Object-Oriented Requirements" of this chapter). An important point to note is that there are many requirements engineering techniques; use case modeling is just one of many.

Yes, use case modeling is important — it is an excellent mechanism to identify and understand the behavioral requirements for your software — but it is not the only technique that you have available to you. Use cases are used as a key input into your planning process, an activity of the Project Management workflow. An important best practice is to prioritize your use cases by both technical and business risk so that you can address the riskiest use cases in the earliest iterations of your project therefore reducing your project's overall risk early in the lifecycle.

Address the riskiest use cases early in your project.

Figure 4.1   The Define and Validate Initial Requirements process pattern.

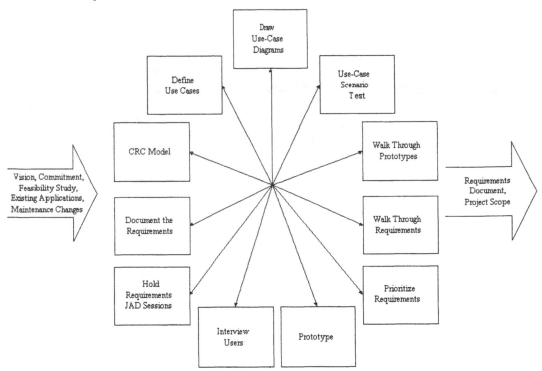

Requirements engineering is an iterative process, as you can see Figure 4.1, comprised of techniques for identifying, documenting, and validating requirements. With an iterative approach, you typically apply the various techniques in any order that you need to, working on a portion of one artifact then switching over to another for awhile. In fact, working on several deliverables simultaneously is a best practice that your organization should consider adopting. It is quite common to find requirements engineers working on a use case model to document behavioral requirements, a Class Responsibility Collaborator (CRC) model to document the key domain entities, and a prototype to understand the requirements for the user interface in parallel. Although this appears to be time consuming on the surface, it actually proves in practice to be very efficient. Because no one artifact is sufficient for all of your requirements engineering needs, you'll find that by focusing on a single artifact, you'll quickly become stuck because you don't know how to use it to document something or its scope is too narrow to fully investigate the problem at hand. Instead, it is more effective to work your use case model for a while and once you run into problems, see if you can address them by hand-drawing some screens or reports. Then, when you run into roadblocks there, you work your CRC model for a bit or move back to your use case model. The Requirements workflow focuses on the identification and documentation of requirements, whereas validation techniques such as walkthroughs and inspections are activities of the Test workflow (Chapter 7).

Working several models in parallel is significantly more productive than focusing on a model at a time.

Another best practice for the Requirements workflow (and for any software development efforts in general) is to document your assumptions as you go. There are several benefits to this. First, by documenting your assumptions, others can put your work into context and will have a greater understanding of what it is you are trying to achieve. Second, by documenting your assumptions, you now have a list of issues that you need to look into — you want to investigate and resolve any assumptions as early as possible. Assumptions are effectively risks to your project, because they are simply guesses upon which you are basing decisions.

Document your assumptions.

A key best practice is that you need to recognize that requirements engineering is a fundamental part of virtually every software development task. If you are working on your organization's business/domain architecture, you should base your efforts on the high-level business requirements captured in your enterprise model. If you are working on your organization's technical architecture, you should base your effort on the technical requirements for your organization, such as the need to support distributed computing in a 24 hours a day, 7 days a week (24/7) manner. If you are evaluating tools, standards, or guidelines for your project team, you should first start by defining the requirements for the items you intend to evaluate (i.e., have a shopping list). If you are designing your system, even if it is highly-technical software such as an operating system or a persistence layer, you should first start by defining the requirements for what you are building. There is a name for developing software without requirements: hacking. Think of it like this: if you don't have any requirements, then you have nothing to build.

Without defined and accepted requirements, you have nothing to build.

4.1 Use Cases and Beyond

In section 4.4.1 "Engineering Object-Oriented Requirements" (*Software Development*, March 1999), I overview the techniques for gathering, documenting, and validating requirements (depicted earlier in Figure 4.1). The article presents two important lessons for software developers. First, there is a wide variety of requirements gathering techniques. Although "use-case driven development" is a nice marketing slogan, what you really want to achieve is a requirements-driven software process. Second, not only do you need to gather and document requirements, you also need to validate them. Experience shows that developers who test often and test early not only produce better software, they are also more efficient doing so. Therefore, validating your requirements via reviews and walkthroughs during the Elaboration phase should be one of your goals. Requirements validation techniques, although covered in this article, are really part of the Test workflow described in Chapter 7. This article also shows how business rules fit into the overall requirements picture, a topic discussed by Colleen McClintock in section 3.3.2 "The Logic of Business Rules" in Chapter 3.

Your goal is to follow a requirements-driven software process.

Not only do you need to understand the techniques for engineering requirements, you also need to understand how to write them well. Karl Wiegers, author of *Software Requirements* (1999), presents sound advice for doing so in section 4.4.2 "Writing Quality Requirements" (*Software Development*, May 1999). In this article, written using the industry-standard terminology of the IEEE, he discusses the best practices for the successful engineering of requirements. He presents a collection of characteristics for quality requirements as well as characteristics of quality software requirements specification (SRS) documents — which the Unified Process refers to as the Requirements Model. Wieger's advice is applicable to any kind of requirements deliverable, including use cases, and is valuable insight that all developers should heed.

Part of requirements engineering is the effective documentation of your software requirements.

In section 4.4.3 "The Various Uses for Use Cases" (*Software Development*, November 1997), I overview the use case modeling process, discussing how you move from use cases to sequence diagrams to class diagrams. As in section 4.4.1 "Engineering Object-Oriented Requirements", I show use cases in the context of other requirements modeling techniques, such as Class Responsibility Collaboration (CRC) cards and your user interface prototype.

It is important to note that in the Spring of 1999 the semantics of use case diagrams evolved — basically the usage of the <<uses>> and <<extends>> stereotypes were reworked. First, the <<uses>> stereotype was renamed <<include>> to make it consistent. It is now singular, with other UML standard stereotypes. More importantly the <<extends>> stereotype (which, as I point out in the article, was used for both generalization and something conceptually similar to a hardware interrupt) was replaced with the generalization relationship (the open-headed arrow) from class modeling and with the <<extend>> stereotype respectively.

To explain the difference, I have reworked the first figure (shown below in Figure 4.2) from the article to use the new modeling notation (show below in Figure 4.3).

Figure 4.2   Use case diagram for a simple business using UML v1.0.

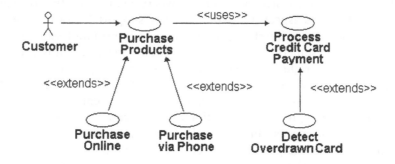

Figure 4.3   Use case diagram for a simple business using UML v1.3.

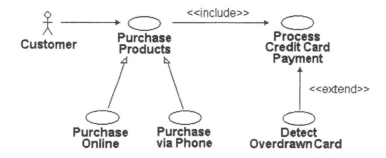

4.2   User Interface Prototyping

For better or worse, the user interface is the system as far as its users are concerned. As a result, users will often focus their efforts on what your HTML pages, screens, and reports look like — other forms of requirements will often be given second priority in favor of user interface development. User interface prototyping focuses on working with users to identify and understand their requirements for your system. As you prototype the user interface, you can find out how they intend to use the system, what the main business concepts are, and any technical requirements they have for the system. This information is key input into your use case model, your CRC model, and your supplementary specifications.

The user interface is the system to your users. Get good at user interface prototyping.

Larry Constantine, co-author of *Software For Use* (Constantine & Lockwood, 1999), presents excellent advice for user interface prototyping in section 4.4.4 "Prototyping From the User's Viewpoint" (*Software Development*, November 1998). As a follow-up to section 3.3.1 "Essentially Speaking" presented in Chapter 3 (where he discusses the concept of understanding the essential business aspects that your software supports), Constantine goes on to show how to perform logical user interface prototyping. An important lesson presented in this article is that just because something looks good, it doesn't mean that it is good. Many software projects lose their way right from the very start when they get distracted by a prototyping tool that produces a slick-looking user interface. During requirements, the goal isn't to design the user interface — it is to understand the requirements for the user interface. That means you don't need to produce a collection of fancy screens; some hand-drawn pictures are often sufficient for your needs. In this article, Constantine presents excellent advice for logical prototyping that enables you to effectively gather the requirements for your user interface. During the Elaboration phase, begin to understand the requirements for your user interface, and during the Construction phase, begin to design it based on those requirements.

It is often a business requirement that your software will be used internationally, therefore your user interface must reflect this fact. In section 4.4.5 "Breaking the Barriers" (*Software Development*, August 1998), Stephen Holmes presents good advice for software developers who need to develop international software. Focusing on the lifecycle issues involved with taking an existing product to a global audience, perhaps via the Internet, Holmes starts with advice for your requirements efforts such as working with an international team and developing the user interface with international issues in mind. Issues such as visual layout, sorting rules for lists of information, double-byte character sets such as Unicode, and language translation all figure prominently in your user interface efforts. Holmes goes on to suggest techniques for internationalizinge your code, including a high-level testing checklist to ensure that your software truly does meet the needs of international users. In fact, this checklist alone provides significant insight into the issues that your requirements efforts must address when developing international software.

There is far more to developing international software than using a double-byte character set such as Unicode.

4.3 Why Are Requirements Important?

Your requirements model should provide the basis for all other aspects of your development efforts. In other words: your software process should be requirements driven. Your requirements define what you are supposed to build, making them key input into your Analysis and Design workflow efforts. In fact, without requirements, you would have nothing to model (unless you're hacking). Your requirements model is also a key input for the definition of your function test cases and your user acceptance test plan which you create as part of your Test workflow efforts. You can also use your requirements model to help you wage the inevitable political battles that all projects must face with other factions within their organization. When another political faction criticizes your work, your architecture and your project plan are typical targets. You can counteract the criticisms by showing how your plan/architecture reflects your requirements as defined by your project stakeholders. This is the application of the *"Requirements As Shield"* process pattern. You can also use your requirements model to support your own political initiatives; the application of the *"Requirements As Stick"* process pattern. The basic idea of this pattern is that whenever another political faction presents an approach — perhaps they wish to use a new development tool or have an alternative architecture to the one that you have formulated — you can use your requirements document to identify flaws in their approach and to justify the importance of these flaws to senior management. As you would imagine, the effective application of both of these process patterns depend on your requirements model being realistic and accepted by your project stakeholders. These patterns are described in detail in volume I: *The Unified Process Inception Phase*.

Your requirements model can be an effective political tool for your project team.

4.4   The Articles

4.4.1 "Engineering Object-Oriented Requirements"
by Scott W. Ambler, March 1999

Understanding the basics of requirements engineering will help you develop better software — and save you time in the long run.

A fundamental reality of our industry is that we develop software based on a set of defined requirements. Perhaps "defined" is the wrong word, as some developers forgo the explicit identification and documentation of requirements. Those developers run into problems more often than not — either their software takes longer to build because of rework, users won't accept it because it doesn't meet their needs, or the project is cancelled outright because upper management is unclear on what value the software will bring to their organization. The gathering and definition of requirements, often referred to as *requirements engineering*, is a key factor for successfully developing software.

All software has requirements. Some developers will argue this — but in my experience, you have nothing to build if you don't have any requirements. Your requirements model is important to the development of your analysis and design models, your project plan, your project charter, and your testing model. Your requirements engineering efforts are the basis from which you perform all other project work.

What should a requirements model contain? First, it should define the software your development team will build. Depending on your situation, you may need to define the functional requirements (oops — I mean, behavioral requirements, since we're talking about objects) so you understand how your users will work with your software. You also need to define the supplementary requirements — such as performance characteristics, operational needs, and the target hardware platform — for your software. If your software will have a user interface, you should develop an initial prototype modeling the screens and reports for it.

Figure 4.4 depicts the Define and Validate Initial Requirements process pattern that depicts the iterative tasks you will perform when engineering object-oriented requirements. I believe you can and should test as early as possible when developing software, which is why there are tasks such as walking through prototypes and use-case scenario testing.

To gather requirements, you can use several techniques. Joint Application Development (JAD) is a structured meeting technique that you can use to define requirements; use-case modeling is a "traditional" object-oriented technique for defining behavioral requirements; Class Responsibility Collaborator (CRC) cards are a user-centered design technique in which users use index cards to actively define their requirements, and, of course, traditional interviewing techniques.

In Figure 4.4, you'll notice that requirements engineering is a complex effort that is performed in an iterative manner. The "star burst" of lines in the diagram's center indicates that you will commonly move back and forth between the tasks you perform. You might gather some requirements, validate them, gather some more, prioritize them, gather some more, validate them, and so on, but you need to be flexible when doing this.

Figure 4.5 shows the potential deliverables of object-oriented requirements engineering. First, note that the Unified Modeling Language (UML), the industry-standard modeling notation for object-oriented development, may not be sufficient for requirements engineering. The UML doesn't include a notation for CRC cards, even though it is a common technique that many CASE tool vendors support, nor does it include a notation for a user interface model. A user interface model includes your user interface prototype, an interface-flow diagram showing the relationships between your screens and reports, and the specifications describing the screens and reports. Luckily, UML collaboration diagrams can serve as a replacement for interface-flow diagrams.

Figure 4.5 also shows that you need to develop several textual deliverables, such as business rules and supplementary specifications, that also aren't defined as part of the UML. Because they still need to be defined, however, they also need to be part of your requirements engineering process. The Rational Unified Process (RUP), the software process that is likely to become the de facto standard for object-oriented development, includes the definition of these deliverables. However, like the UML, the Rational Unified Process falls short with respect to the user interface model (of course, I'm being facetious by suggesting that when the folks at Rational purchase a prototyping tool vendor, perhaps the importance of this issue will become clear to them).

Figure 4.4   Defining and Validating Initial Requirements.

Figure 4.5   Defining object-oriented requirements.

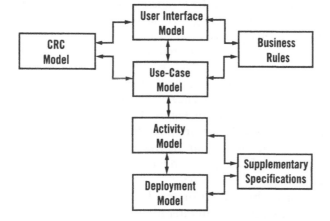

Now that you understand the engineering requirements process, and the deliverables you will produce in that process, how does requirements engineering fit into the rest of your software development efforts? Figure 4.6 depicts the Detailed Modeling process pattern that shows the relationships between the main deliverables of object-oriented modeling, indicating how the initial deliverables of your requirements engineering efforts fit into your overall development efforts. The information contained in your use cases drives the development of sequence diagrams, showing the detailed logic of one scenario, which in turn drives the development of your class models. Your use cases also affect your class model, and vice versa, because business classes (such as Account and Customer in a banking environment) are likely to appear in both deliverables — as you develop one deliverable you will discover information that pertains to the other.

Figure 4.6   Detailed Modeling.

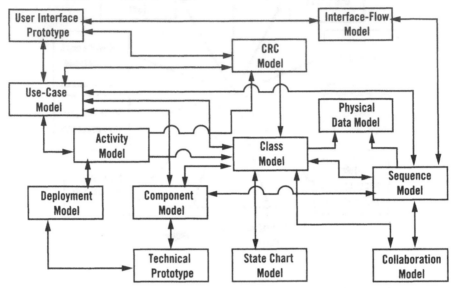

It is also interesting to note that the business rules and supplementary specifications identified during requirements engineering also drive the development of your object-oriented models. For example, a business rule — such as requiring a $500 minimum balance in a checking account — would be implemented in one or more classes in your class model. A technical requirement that appears in your supplementary specifications, such as requiring a deposit be processed in less than one second, would likely drive features documented in one or more state chart diagrams and sequence diagrams.

We can learn several lessons from Figure 4.6. First, the very nature of object-oriented requirements engineering, and object-oriented development in general, suggests the need to maintain traceability throughout your deliverables. Luckily, the deliverables apply basic object-oriented concepts such as classes, collaboration, and relationships in a consistent manner, making traceability significantly easier.

The second lesson is that object-oriented development isn't completely iterative — I believe it is serial in the large-scale and iterative in the small-scale. By this I mean that at the beginning of development, you spend most of your time doing requirements tasks, then you move into analysis, then design, then implementation, then testing. You can see this in Figure 4.6 where the deliverables in the top-left corner are mostly requirements deliverables. Moving toward the bottom-right are analysis deliverables, then design deliverables. If Figure 4.6 showed source code, you would see that it is driven by your class model, sequence diagrams, state chart diagrams, and collaboration diagrams. I call this serial in the large-scale. On any given day, however, you are likely to do some requirements, analysis, design, and even coding. I call this iterative in the small-scale. In short, the idea that object development is purely iterative is questionable.

The Potential Deliverables of Object-Oriented Requirements Gathering

Activity model — One or more UML activity diagrams and the corresponding documentation that describes the activity and states what they model and the transitions between them. Activity models describe a system's dynamic nature and are often used to document the underlying business process.

Business rule — A prescription of an action or guideline that your system must support. Text and a supporting activity diagram often describe business rules.

CRC model — A Class Responsibility Collaborator (CRC) model is a collection of CRC cards. A CRC card represents a single class, listing its responsibilities, what it knows and does, and the class collaborators, which are the other classes it works with to fulfill its responsibilities.

Deployment model — One or more UML deployment diagrams and the corresponding documentation describing the nodes within your system and the dependencies and connections between them. Deployment models describe your system's static hardware and software configuration.

Supplementary specification — A collection of non-behavioral requirements, including regulatory requirements, quality requirements (performance, reliability, supportability, and so on), and technical requirements, such as definition of the operational platform.

Use-case model — One or more use-case diagrams and the corresponding documentation describing the actors, use cases, and the relationships between them for your system. Use-case models describe your system's functional or behavioral requirements.

User interface model — The collection of user interface artifacts, such as the user interface prototype, the interface-flow diagram describing the relationships between screens and reports, and the supporting documentation describing each screen and report within your system.

In this column, I explored some of the basic techniques and potential deliverables of object-oriented requirements engineering, many of which are defined by industry-standard UML. In future columns, I will explore several requirements engineering techniques in greater detail.

4.4.2 "Writing Quality Requirements"

by Karl Wiegers, May 1999

All too often, software requirements are badly written and hard to follow. Clarifying your specifications will benefit everyone involved.

It looks like your project is off to a good start. Your team had customers involved in the requirements elicitation stage and you actually wrote a software requirements specification. The specification was big, but the customers signed off on it, so it must have been O.K.

Now you're designing one of the features and you've found some problems with the requirements: You can interpret requirement 15 a couple different ways. Requirement 9 states precisely the opposite of requirement 21; which should you believe? Requirement 24 is so vague that you don't have a clue what it means. You just had an hour-long discussion with two other developers about requirement 30 because all three of you thought it meant something different. And the only customer who can clarify these points won't return your calls. You're forced to guess what many of the requirements mean, and you can count on doing a lot of rework if you guess wrong.

Many software requirements specifications (SRS) are filled with badly written requirements. Because the quality of any product depends on the quality of its raw materials, poor requirements cannot lead to excellent software. Sadly, few software developers have been educated about how to elicit, analyze, document, and verify requirement quality. There aren't many examples of good requirements to learn from, partly because few projects have good ones to share, and partly because few companies are willing to place their product specifications in the public domain.

This article describes several characteristics of high-quality software requirement statements and specifications. I will examine some less-than-perfect requirements from these perspectives and take a stab at rewriting them. I've also included some general tips on how to write good requirements. Evaluate your project's requirements against these quality criteria. It may be too late to revise them, but you might learn how to help your team write better requirements next time.

Don't expect to create a SRS in which every requirement exhibits every desired characteristic. No matter how much you scrub, analyze, review, and refine requirements, they will never be perfect. However, if you keep these characteristics in mind, you will produce better requirements documents and you will build better products.

Guidelines for Writing Quality Requirements

There is no simple formula for writing excellent requirements. It is largely a matter of experience and learning from past requirements problems. Here are a few guidelines to keep in mind as you document software requirements.

- Keep sentences and paragraphs short. Use the active voice. Use proper grammar, spelling, and punctuation. Use terms consistently and define them in a glossary.

- To see if a requirement statement is sufficiently well-defined, read it from the developer's perspective. Mentally add the phrase, "call me when you're done" to the end of the requirement and see if that makes you nervous. In other words, would you need additional clarification from the SRS author to understand the requirement well enough to design and implement it? If so, elaborate on that requirement before you proceed.

- Requirement authors often struggle to find the right level of granularity. Avoid long narrative paragraphs that contain multiple requirements. Write individually testable requirements. If you can think of a small number of related tests toverify a requirement's implementation, it's probably written at the right level of detail. If you envision many different kinds of tests, perhaps several requirements have been lumped together and should be separated.

- Watch out for multiple requirements that have been aggregated into a single statement. Conjunctions like "and" and "or" in a requirement suggest that several requirements have been combined. Never use "and/or" or "etc." ina requirement statement.

- Write requirements at a consistent level of detail throughout the document. I have seen requirements specifications that varied widely in their scope. For example, "A valid color code shall be R for red" and "A valid color code shall be G for green" might be split into separate requirements, while "The product shall respond to editing directives entered by voice" describes an entire subsystem, not a single functional requirement.

- Avoid stating requirements redundantly in the SRS. While including the same requirement in multiple places may make the document easier to read, it also makes it more difficult to maintain. You must update multiple instances of the requirement at the same time to avoid inconsistency.

Characteristics of Quality Requirement Statements

How can you distinguish good software requirements from problematic ones? Individual requirement statements should exhibit six characteristics. A formal inspection of the SRS by project stakeholders with different perspectives is one way to determine whether or not each requirement has these desired attributes. Another powerful quality technique is writing test cases against the requirements before you cut a single line of code. Test cases crystallize your vision of the product's behavior as specified in the requirements and can reveal omissions and ambiguities.

Characteristic #1: They must be correct.   Each requirement must accurately describe the functionality to be delivered. The reference for correctness is the source of the requirement, such as an actual customer or a higher-level system requirements specification. A software requirement that conflicts with a corresponding system requirement is not correct (of course, the system specification could be incorrect, too).

Only users can determine whether the user requirements are correct, which is why it's essential to include actual users, or surrogate users, in requirements inspections. Requirements inspections that do not involve users can lead to developers saying, "That doesn't make sense. This is probably what they meant." This is also known as "guessing."

Characteristic #2: They must be feasible. You must be able to implement each requirement within the known capabilities and limitations of the system and its environment. To avoid infeasible requirements, have a developer work with the requirements analysts or marketing personnel throughout the elicitation process. This developer can provide a reality check on what can and cannot be done technically, and what can be done only at excessive cost or with other trade-offs.

Characteristic #3: They must be necessary for the project. Each requirement should document something the customers need or something you need to conform to an external requirement, an external interface, or a standard. You can think of "necessary" as meaning each requirement originated from a source you know has the authority to specify requirements. Trace each requirement back to its origin, such as a use case, system requirement, regulation, or some other voice-of-the-customer input. If you cannot identify the origin, perhaps the requirement is an example of gold-plating and isn't really necessary.

Characteristic #4: They must be prioritized. Assign an implementation priority to each requirement, feature, or use case to indicate how essential it is to a particular product release. Customers or their surrogates have the lion's share of the responsibility for establishing priorities. If all the unprioritized requirements are equally important, so is the project manager's ability to react to new requirements added during development, budget cuts, schedule overruns, or a team member's departure. You can determine priority by considering the requirement's value to the customer, the relative implementation cost, and the relative technical risk of implementing it.

Many groups use three levels of priority. High priority means you must incorporate the requirement in the next product release. Medium priority means the requirement is necessary but you can defer it to a later release if necessary. Low priority means it would be nice to have, but you might have to drop it because of insufficient time or resources.

Characteristic #5: They must be unambiguous. The reader of a requirement statement should draw only one interpretation of it. Also, multiple readers of a requirement should arrive at the same interpretation. Natural language is highly prone to ambiguity. Avoid subjective terms like user-friendly, easy, simple, rapid, efficient, several, state-of-the-art, improved, maximize, and minimize. Write each requirement in succinct, simple, straightforward language of the user domain, not in technical jargon. You can reveal ambiguity through formal requirements specifications inspections, writing test cases from requirements, and creating user scenarios that illustrate the expected behavior of a specific portion of the product.

Characteristic #6: They must be verifiable. See whether you can devise tests or use other verification approaches, such as inspection or demonstration, to verify that the product properly implements each requirement. If you can't verify a requirement, determining whether or not it was correctly implemented is a matter of opinion. Requirements that aren't consistent, feasible, or unambiguous are also not verifiable. Any requirement that the product shall "support" something isn't verifiable.

(Discourse continues on page 110.)

Reviewing Quality Requirements

These descriptions of quality requirements characteristics are fine in the abstract, but what do good requirements really look like? To make these concepts more tangible, review the following examples. Each is a requirement adapted from an actual project. Evaluate each one against this article's requirements quality criteria, and rewrite it in a better way. I'll offer my analysis and improvement suggestions for each, but you might come up with a different interpretation. I have an advantage because I know where each requirement came from. Since you and I aren't the real customers, we can only guess each requirement's actual intent.

Example #1: "The product shall provide status messages at regular intervals not less than every 60 seconds." This requirement is incomplete. What are the status messages and how are they supposed to be displayed to the user? It also contains several ambiguities. What part of "the product" does it refer to? If the interval between status messages is really at least 60 seconds, then is showing a new message every 10 years acceptable? Perhaps the intent is to have no more than 60 seconds elapse between messages; then would 1 millisecond be too short? The word "every" just confuses the issue. As a result of these problems, the requirement is not verifiable.

Here is one way you could rewrite the requirement to address those shortcomings:

1. Status Messages.
   - 1.1 The Background Task Manager shall display status messages in a designated area of the user interface at intervals of 60 plus or minus 10 seconds.
   - 1.2. If background task processing is progressing normally, the percentage of the background task processing that has been completed shall be displayed.
   - 1.3. A message shall be displayed when the background task is completed.
   - 1.4. An error message shall be displayed if the background task has stalled.

I split this into multiple requirements because each will require separate test cases and should be separately traceable. If several requirements are strung together in a paragraph, it's easy to overlook one during construction or testing.

Example #2: "The product shall switch between displaying and hiding non-printing characters instantaneously." Computers cannot do anything instantaneously, so this requirement is not feasible. It is incomplete because it does not state what conditions trigger the state switch. Does the software make the change on its own under certain conditions, or does the user take some action to stimulate the change? Also, what is the scope of the display change within the document: selected text, the entire document, or something else? There is an ambiguity problem, too. Are "non-printing" characters the same as hidden text, or are they attribute tags or control characters of some kind? Because of these problems, this requirement cannot be verified.

This might be a better way to write the requirement: "The user shall be able to toggle between displaying and hiding all HTML markup tags in the document being edited with the activation of a specific triggering condition." Now it is clear that the non-printing characters are HTML markup tags. This requirement does not constrain the design because it does not define the triggering condition. When the designer selects an appropriate triggering condition, you can write specific tests to verify that this toggle operates correctly.

Example #3: "The HTML parser shall produce an HTML markup error report that allows quick resolution of errors when used by HTML novices." The word "quick" is ambiguous. The lack of definition of what goes into the error report is a sign of incompleteness. I'm not sure how you would verify this requirement. Would you find an HTML novice and see if he or she can resolve errors quickly enough using the report?

Try this instead: "The HTML parser shall produce an error report that contains the line number and text of any HTML errors found in the parsed file and a description of each error found. If no errors are found, the error report shall not be produced." Now you know what needs to go into the error report, but you've left it up to the designer to decide what the report should look like. You have also specified an exception condition: if there aren't any errors, don't generate a report.

Example #4: "Charge numbers should be validated online against the master corporate charge number list, if possible." I give up, what does "if possible" mean? If it's technically feasible? If the master charge number list can be accessed online? Avoid imprecise words like "should." The customer either needs this functionality or doesn't. I've seen some requirements specifications that draw subtle distinctions between key words like "shall," "will," and "should" to indicate priority. I prefer to use "shall" as a clear statement of what I intend by the requirement and to explicitly specify the priorities.

Here is an improved version of this requirement: "The system shall validate the charge number entered against the online master corporate charge number list. If the charge number is not found on the list, an error message shall be displayed and the order shall not be accepted."

Characteristics of Quality Requirements Specifications

A complete SRS is more than a long list of functional requirements. It also includes external interface descriptions and nonfunctional requirements, such as quality attributes and performance expectations. Look for the following characteristics of a high-quality SRS.

Characteristic #1: It is complete. The SRS shouldn't be missing any requirements or necessary information. Individual requirements should also be complete. It is hard to spot missing requirements, because they aren't there. Organize your SRS's requirements hierarchically to help reviewers understand the structure of the described functionality. This makes it easier to tell if something is missing.

If you focus on user tasks rather than on system functions when gathering requirements, you're less likely to overlook requirements or include requirements that aren't really necessary. The use case method works well for this purpose. Graphical analysis models that represent different views of the requirements will also reveal incompleteness.

If you know you are lacking certain information, use "TBD" (to be determined) as a standard flag to highlight these gaps. Resolve all TBDs before you construct the product.

Characteristic #2: It is consistent. Consistent requirements do not conflict with other software requirements or with higher level (system or business) requirements. You must resolve disagreements before you can proceed with development. You may not know which (if any) is correct until you do some research it. Be careful when modifying the requirements; inconsistencies can slip in undetected if you review only the specific change and not related requirements.

Characteristic #3: It is modifiable. You must be able to revise the SRS when necessary and maintain a history of changes made to each requirement. You must give each requirement a unique label and express it separately from other requirements, so you can refer to it clearly. You can make a SRS more modifiable by grouping related requirements, and by creating a table of contents, index, and cross-reference listing.

Characteristic #4: It is traceable. You should be able to link each software requirement to its source, which could be a higher-level system requirement, a use case, or a voice-of-the-customer statement. Also, link each software requirement to the design elements, source code, and test cases that implement and verify it. Traceable requirements are uniquely labeled and written in a structured, fine-grained way. Bullet lists are not traceable.

The difficulty developers will have understanding poorly written requirements is a strong argument for having both developers and customers review requirements documents before they are approved. Detailed inspection of large requirements documents is not fun, but it's worth every minute you spend. It's much cheaper to fix the defects at that stage than later in the development process or when the customer calls.

If you observe the guidelines for writing quality requirements (see sidebar) and if you review the requirements formally and informally, early and often, your requirements will provide a better foundation for product construction, system testing, and ultimate customer satisfaction. And remember that without high-quality requirements, software is like a box of chocolates: you never know what you're going to get.

4.4.3 "The Various Uses for Use Cases"

by Scott W. Ambler, November 1997

Use cases and use case diagrams are tools used for gathering, defining, and validating user requirements. You can use them both to model your enterprise or for a specific business application.

This month I will take a look at an object-oriented modeling technique that you've probably heard a lot about: use cases. The interesting thing is that although everyone focuses on use cases, the actual secret to success is to understand use case diagrams and how they fit into the overall modeling process. Before I begin, I'd like to point out that you can implement use case diagrams on any project, whether it's object-oriented or not; however, because use cases focus on behavior, they fit extremely well with the object-oriented paradigm.

Use case diagrams show several things: use cases, which are a series of related transactions that provide an important behavior; actors, which interact with your organization and system; and the associations between the actors and uses cases within the diagram. Figure 4.7 illustrates how a simple business can follow the Unified Modeling Language (UML v1.0) notation for use case diagrams. Use cases are shown as horizontal ovals, actors are shown as stick figures, and associations are shown as arrows. Figure 4.3 depicts this figure in UML v1.3 notation.

A use case describes a series of steps to be taken that fulfill a fundamental behavior of a business or application. For example, the Purchasing Products use case in Figure 4.7 has the steps total items, calculate applicable discount, calculate applicable taxes, calculate grand total, charge customer, make change for customer, thank customer, and give customer purchased products.

As you can imagine, there are several different ways a customer can purchase products. What I just described is the most common way for customers to purchase products, but it's not the only way. That's one reason why we have the "extends association:" to effectively redefine part or all of an existing use case. In Figure 4.7, we see that the use cases Purchasing Online and Purchasing via Phone both extend Purchasing Products, redefining the necessary steps that are specific to those versions of purchasing. In many ways, this form of extends is the use case equivalent of inheritance because the extending use cases override some or all of the behavior in the original use case.

Figure 4.7   A use case diagram for a simple business.

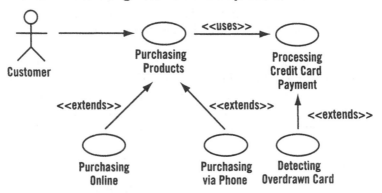

But what about the other extends association shown in Figure 4.7? The use case Detecting Overdrawn Card extends Processing Credit Card Payment in a slightly different way — instead of redefining the extended use case, it interrupts it. Detecting Overdrawn Card, effectively an error condition use case that interrupts the Processing Credit Card Payment use case, marks the card so that it can no longer be used until a payment is made. Then it returns control to the original use case, which continues to process as before. The assumption is that the payment still goes through even though the card is now overdrawn, but no further transaction will be accepted until a payment is made.

Figure 4.7 also shows a "uses association" from Purchasing Products to Processing Credit Card Payment, indicating that at some point during the execution of the Purchasing Products use case, the credit card use case is invoked. Why wouldn't you simply include Processing Credit Card Payment in Purchasing Products? First, the processing of credit card payments is a complex and cohesive thing, so it makes sense to consider creating a use case for it. Second, and more important, credit card processing is a common service that can be used by other use cases, perhaps use cases like Reserving Room and Charging Fines, because your organization does more than just sell products to customers. When you stop and think about it, perhaps a better name for the uses association would have been the "reuses" association because it promotes reuse within your use case diagram.

(Discourse continues on page 114.)

Common Use Case Terminology:

Actor   An actor defines one or more roles that someone or something in the environment can play in relation to the business or system. An actor is anything that interacts (exchanges data or events) with the system or business. Actors are shown on use case diagrams as stick figures.

Class Diagram   Class diagrams show the classes of the system, their interrelationships, and the collaborations between those classes. Class diagrams are often (mistakenly) referred to as object models.

Component Diagram   A diagram that shows the software components, their interrelationships, interactions, and public interfaces that comprise an application, system, or enterprise.

Class Responsibility Collaborator Model   A collection of standard index cards divided into three sections showing the class's name, responsibilities, and collaborators.

Enterprise Modeling   The act of modeling the business, or problem domain, of an entire organization. The focus is on defining the high-level business behaviors of the organization and not the implementation of those behaviors. The enterprise model defines the context in which packages exist.

Extends Association   An indication that one use case potentially enhances another according to a given condition. If use case B extends use case A, at some point in time an object obeying use case A may temporarily discontinue doing so in favor of obeying use case B. Once use case B completes its action, the object continues obeying use case A where it originally left off. From a hardware perspective, we would say that use case B interrupts use case A. An extends association is shown on a use case diagram by an arrow labeled <<extends>>.

External Actor   An actor that is outside of your organization. External actors are typically other organizations, people, or systems owned and operated by other organizations.

Internal Actor   An actor that is within your organization. Internal actors are typically a position or another system within the organization.

Sequence Diagram   A diagram that shows the types of objects involved in a use case scenario, including the messages they send to one another and the values that they return.

Use Case   A sequence of transactions that yield a result of measurable value to an actor or to another use case. Use cases are shown on use case diagrams as horizontal ovals.

Use Case Diagram   A diagram showing a set of use cases, actors, and their associations.

Use Case Scenario   A specific example of a use case, often referred to as an instance of a use case.

Uses Association   A relationship defining a use case description that uses another use case description as part of its own. A uses association is shown on a use case diagram by an arrow labeled <<uses>>.

How Use Case Diagrams Fit In

Use cases and use case diagrams are tools for gathering, defining, and validating user requirements. You can use them to define the high-level business requirements for your enterprise, and to define the detailed requirements for a specific application. Figure 4.8 shows how use cases fit into enterprise modeling, while Figure 4.9 shows how use cases fit into detailed application modeling.

Figure 4.8   Enterprise business modeling.

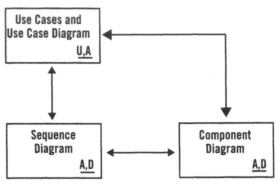

A=Analyst, D=Designer, U=User

In Figure 4.8, three main diagrams make up the enterprise model: use case diagrams, sequence diagrams, and component diagrams. The boxes represent object-oriented modeling techniques and the arrows show the relationship between them — with the arrow heads indicating an "input into" relationship. In the bottom right-hand corner of each box are letters indicating who is typically involved on that diagram or technique: "U" for user, "A" for analyst, and "D" for designer. The letter underlined indicates the individual or group performing the majority of work for that diagram. As you would expect with iterative development, the information contained in each diagram helps to drive the development and enhancement of the other diagrams. Having said that, enterprise modeling typically begins with use case modeling, followed by development of the other two diagrams, which are used to verify and complete the original use cases.

Use case diagrams and their corresponding use cases are the main artifacts created during enterprise modeling. Working closely with a group of business domain experts, also called subject matter experts, an experienced use case modeler can develop a set of high-level use cases that describe what the business does. With enterprise modeling, the use cases describe what the entire business does, they do not describe an information system that only supports a portion of the business. That's what detailed analysis is all about.

Figure 4.9 shows how use case diagrams are used for iteratively developing a detailed analysis model for an object-oriented application. Once again, modelers work closely with business domain experts to develop the model, except this time the focus is on a specific application instead of the entire business. Application use cases are typically much more detailed than enterprise-level use cases because they document the behavioral requirements for your application. It is also quite common to develop a class responsibility collaborator

(CRC) model and a series of user interface prototypes simultaneously with the use case diagram. This is because all three of these techniques require you to work closely with your users.

Figure 4.9   Use cases in detailed application analysis.

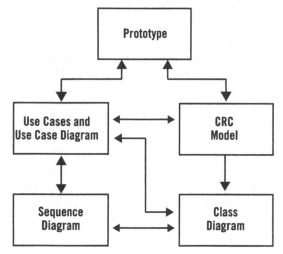

Use cases are an effective way to understand and document your problem domain. Use cases and use case diagrams are straightforward modeling techniques that you can use when working closely with your users to understand both the enterprise in which you work and specific applications. They also fit in nicely with other object-oriented modeling techniques, such as sequence diagrams and class diagrams, providing you with a valuable technique for gathering the domain information that is critical to your project's success.

4.4.4 "Prototyping From the User's Viewpoint"
by Larry Constantine, November 1998

Easy-to-use visual development tools can actually hide or neglect the aspects of the user interface that most profoundly affect usability. Abstract prototyping can help.

The programmer smiled proudly as he handed us a floppy disk. It was the third day of a week-long class on usage-centered design, and while the rest of the class was working on design models and paper prototypes, this enterprising developer had hacked up a visual prototype in Delphi on his laptop. We loaded the floppy and launched the application. Compared to the collections of sticky notes and rough pencil sketches scattered over tabletops around the room, his prototype looked a lot prettier and more like a real program. That, precisely, was the problem. Its good looks hid numerous design problems that would have been easier to spot and avoid if the prototype had not looked quite so real or so concrete.

Abstract prototyping is a way to avoid the seduction of attractive prototypes that disguise weak designs. By using sticky notes or other simple tricks, abstract prototypes can represent the contents of a user interface without showing what it looks like. Lucy Lockwood and I developed this design technique to bridge the gap between user problems and visual prototypes. By making better use of modern visual development tools, abstract prototyping can speed and simplify the design of highly usable systems and help you produce improved, more innovative software products.

Trapped in Concrete

With the advent of modern visual development tools such as Delphi, Visual Studio, JBuilder, and VisualAge, constructing GUIs has become so easy there hardly seems to be any need for design at all. Just open a new form, throw on a data-aware display grid, drag-and-drop a menu bar plus a few widgets from a tool palette, and without so much as a pause for a sip of coffee, you have the user interface all laid out. And it looks pretty good, too!

On the plus side, visual development tools facilitate visual thinking, making it easy to lay out user interfaces merely by fiddling around and eyeballing the results. If you have used such systems, you don't need to be told that they're fast, powerful, and easy to use. However, their very simplicity can encourage rapid and undisciplined assembly of user interfaces. You can create forms, windows, or dialog boxes so easily that interaction contexts can multiply like tribbles. Further, the assortment of GUI widgets standing ready for immediate instantiation encourages the proliferation of features. The result can be bloated interfaces, stuffed with features, that take the user on a merry chase from dialog to dialog.

The power of the tools reinforces the practice of just throwing something together in the hope of fixing it later. But if there isn't enough time to think it through and do it right today, there probably won't be enough time later. Usually, all that is accomplished later is a bit of aesthetic polish, straightening out the alignment of some of the widgets, adding a glyph here or there, and reordering a few controls.

Because visual development tools can make even careless designs look good, those aspects of user interfaces that most profoundly affect usability may end up being neglected. What capabilities must be present within the user interface for it to solve the users' problems? How should they be organized into handy collections? How should the work flow within and between the various parts of the interface? You can easily answer all these questions with abstract models.

Why Model?

As a breed, visual developers don't spend a lot of time building models. Like the participant in our design training, many would probably argue that it's quicker and easier just to start programming.

We use models in our work because they actually speed up development and help us find a superior solution more quickly. Good models clarify design issues and highlight trade-offs, so you can resolve decisions rapidly. Perhaps most important, models help us know what to build so we don't waste time building the wrong system or creating unnecessary features. Even under the tightest deadlines, the most effective developers construct analysis and design models because they know that modeling actually saves time. In fact, modeling may become even more vital as the timescale of development is compressed.

Models also help you deliver better, more robust systems. You wouldn't want to fly in a commercial aircraft that had been built without benefit of wind tunnel tests of scale models or simulations using computer models. As in other fields, models help software developers document the results of discussions and communicate their understanding of the issues and answers to other members of the development team, to clients, and to users.

Lucy and I devised abstract prototyping because we found that the sooner developers started drawing realistic pictures or positioning real widgets, the longer it took them to converge on a good design. Abstract models are always much simpler than the real thing. They help you avoid being misled or confused by messy and distracting detail, making it easier for you to focus on the whole picture and overall issues. Concrete tools tend to lead to concrete thinking. They draw you into thinking in conventional and familiar terms and hinder you from thinking creatively. Abstract models invite innovation, encouraging developers to fill in the details more creatively.

Content Modeling

Content models are a tool for abstract prototyping. A content model represents the contents of a user interface without regard for the details of what the user interface looks like or how it behaves. In a content model, a user interface is a collection of materials, tools, and working spaces, all of which are described in terms of the functions they serve, their purposes, or how they are used. Materials are simply what users are interested in seeing or manipulating, tools are what enable users to do things with materials, and working spaces are the various parts of the user interface that combine the tools and materials into useful collections.

Good user interface design assures that all the tools and materials the user needs are on the user interface and are assembled into collections that are convenient for the tasks to be performed with the software.

It is not hard to see that materials, tools, and working spaces can be used to describe a lot more than just software user interfaces. Most tasks are carried out in particular places using appropriate collections of tools and materials to complete the work. A machine shop and a kitchen are laid out differently to suit different tasks. They contain different equipment and supplies, which are stored and organized in ways suited to their distinct uses. Before baking a lemon poppy-seed cake, I will assemble a set of utensils and ingredients that are different than those I'd use while cooking up a potato curry.

Good user interface design, then, is a matter of assuring that all the tools and materials the user needs are found on the user interface and are assembled into collections that are convenient for the tasks to be performed with the software. That is what abstract prototyping can help you achieve.

Low-Tech, Low-Fi

We take a decidedly low-tech approach to abstract prototyping. In fact, in our experience, better designs result from more primitive, and therefore more abstract, modeling tools.

We use sheets of paper to represent the working spaces (or interaction contexts, as they are sometimes called) within the user interface, and fill them with sticky notes to represent the needed tools and materials: hot colors for tools and cool colors for materials. Sticky notes have a lot of advantages for this purpose. They are easily rearranged, so you can quickly try a variety of radically different organizations. Sticky notes also do not look much like real GUI widgets. You can invent or invoke anything you can describe or define in words without worrying about what it will look like or how you'll program it. Some human-computer interaction specialists refer to this sort of a model as a "low-fidelity prototype," because it doesn't look like a real user interface. This view misses the point: a content model is really a high-fidelity abstract model.

To illustrate content modeling, let's explore a highly simplified example. Imagine you're designing a tool that software developers will use for tracking usability problems and user interface defects in the software they produce. You want a simple way to mark the location of each problem on the user interface design to identify the type of problem, annotate it with a brief description if needed, and indicate its seriousness. For instance, you might want to mark a tool button and note that it has an obscure icon — a moderately serious problem. You may know from experience that usability problems fall into predictable categories, which no one wants to keep re-entering. Predefined problem types will also make it easier to track performance over time, learning what mistakes to avoid. You'll also want to track the status of identified defects along with the responsible developer.

Essential Use Cases

Usage-centered design employs a simplified, abstract form of use cases known as essential use cases. Rather than being cast as concrete and specific user actions and system responses, as are the more traditional or concrete use cases, essential use cases are expressed in terms of user intentions and system responsibilities. For example, for recordingUsabilityDefect, the essential use case might have a narrative like this:

recordingUsabilityDefect

USER INTENTION	SYSTEM RESPONSIBILITY
optionally select interface context	show interface context
indicate defect location	show marked interface context suggest defect type
pick type of defect optionally describe note defect severity	show entered details
confirm	remember defect with details

Insofar as practical, essential models are technology- and implementation-independent. The essential use case shown here doesn't mention how information is displayed or how selections are made, for example. Such use cases provide a smooth transition into abstract prototyping using content models.

Starting Points

How do you start the content model? Developers who are most comfortable thinking in terms of procedures or methods may be inclined to start thinking about the tools or active features of the user interface, but we find we get better results by figuring out first how to meet users' information needs. Once you meet the data and information requirements, you can easily identify the tools and functions needed to manipulate or transform these materials. Others, including S. Lauesen and M.B. Harning in their chapter on "Dialogue Design Through Modified Dataflow and Data Modeling" (Human-Computer Interaction, Springer-Verlag, 1993), have used similar approaches and have also found advantages in giving first attention to the data or information required within the user interface.

You can begin by asking simple questions: What information will the user need to create, examine, or change? What information will the user need to present or offer to the system? What information will the user require from the system? What views, perspectives, or forms of presentation will be of value to the user? What conditions or events will the user need to convey to the system, and about what conditions or events will the user need to be informed?

If you're familiar with use cases, you will recognize such questions, which are mostly the same questions asked in identifying use cases. This is no accident. One goal of good user interface design is to provide simple and direct support of the tasks of interest to users. In usage-centered design, the interface content model is derived from essential use cases (see sidebar), which makes the process more straightforward and favors clean, simple designs. However, even if you're not working from use cases, thinking in terms of basic user needs can still be effective.

You can start by labeling a piece of paper as the current working space. The working space could become an application window, a dialog box, a child window, or just one page in a tabbed dialog, but you should defer the choice until later. At this point, the main issue is what spaces will contain what tools and materials. The name you give to a working space usually reflects what users will do in that part of the user interface. Let's call this particular working space, "Recording Usability Problems."

To record usability problems, the user will need some kind of picture of each part of the user interface, each interaction context within the design being tracked. Since a complete user interface design typically consists of a collection of dialogs, forms, and other contexts, the user will need access to the entire collection. So, you can label some sticky notes for these materials — an `interfaceContextHolder` to hold an image, an `interfaceContextName`, and an `interfaceContextCollection` — and slap them onto the paper. To ensure standard descriptions of usability problems, you can add a collection of `usabilityDefectTypes` from which the user can select. Since you can think of a specific usability problem as consisting of a type, a description, and a degree of severity, you can simplify the picture by combining these closely related ideas into a single material called the `usabilityProblem`.

Materials need to be supported with appropriate tools. To identify the necessary tools, you can ask questions like: How will users need to operate upon or change the information of interest on the user interface? What capabilities or facilities will need to be supplied by the user interface for users to be able to accomplish their tasks?

For the usability problem tracker, you need to be able to switch images of interaction contexts. Thus, you can add a `contextNavigator` tool to page forward and back through a collection of images. If you think carefully about how the defect tracker will be used, you'll realize that the user will sometimes need to switch directly from one image to another. Thus, you can add an `interfaceContextSelector` tool. Since you'll need a means for marking the location of problems or defects, you can also add a `defectLocationMarker` tool. As you add abstract components, you might have some ideas about their implemented form; these you can write directly on the sticky notes, not as final decisions but as possible suggestions.

At this stage, you might have something like the working space that is illustrated in Figure 4.10. As shown, it can be useful to group tools and materials together to indicate their relationships or suggest similarities.

Figure 4.10   Working space: recording usability problems.

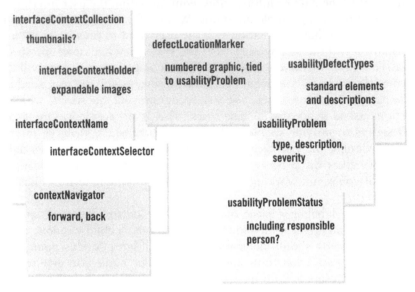

Figure 4.11   A simple notation for the navigation map.

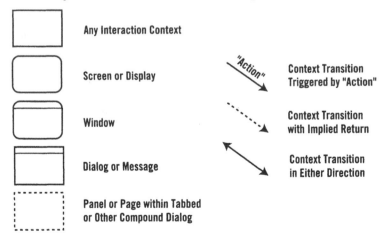

Is this enough to solve the problem of recording usability defects? It's hard to be certain without knowing everything the user will want to do. Certainly for a complete, practical system, you'd need more. You'd need some housekeeping or administrative tools, such as for saving an annotated collection of interface images. In addition, the user would need to see and print lists of all the defects in a project, summarized by interaction context, defect type, status, and responsible person, for example.

If you started creating a new working space for "Reviewing Usability Defects by Interaction Context," you would quickly find yourself filling it up with almost all the same tools and materials as the "Recording Usability Problems" working space. You would not need a usabilityProblemTypes, but you would need an entire list of problem descriptions and a whole set of problem markers. In content modeling, whenever you see two working spaces with substantially the same contents, you will want to consider combining them if it will simplify the user interface. In this case, merging the two working spaces into one probably has more advantages than disadvantages.

Fairly quickly, you will have identified several distinct but interrelated working spaces or interaction contexts. How these are interconnected is an important part of the overall organization — the architecture — of the user interface. A complete content model consists of two parts: the content model itself, and a navigation map that shows how all the different interaction contexts are interconnected so the user can navigate among them. In practice, you can develop these two views of the interface contents in tandem.

A simple and flexible notation suffices for the navigation map, as shown in Figure 4.11. The navigation map lets you look at the complete organization of even complex user interfaces with many screens or dialog boxes. Of course, the usability defect tracker is not very complex; Figure 4.12 represents one possible navigation map for it.

Figure 4.12   Partial navigation map.

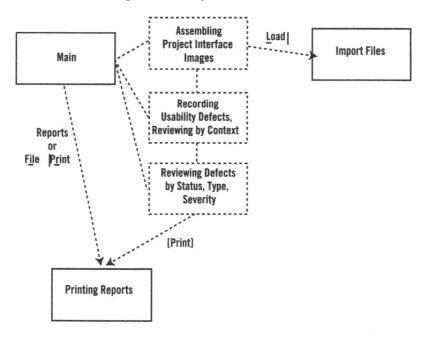

Down to Specifics

For most designers, translating an abstract prototype into a visual design or a working prototype is where the real fun begins. Typically, you would start with a somewhat naïve transliter-
ation: one abstract component implemented as a single standard GUI widget. For example,
the interfaceContextHolder could become a simple bitmap display that expands to full
screen on double-click.

Often, however, there are opportunities for creative combinations that save screen real
estate or simplify operation. For example, you could implement the interfaceContextName
and interfaceContext Selector as separate widgets, but a single drop-down combo box
would cover both functions. Alternatively, you could combine the forward-and-back functions of the contextNavigator with interfaceContextName using a spin box. All three capabilities might be realized with the single custom widget shown in Figure 4.13. Although
non-standard, carefully crafted components based on simple visual and behavioral theme and
variation can be easy for users to figure out and can significantly improve usability.

Figure 4.13   A single custom widget.

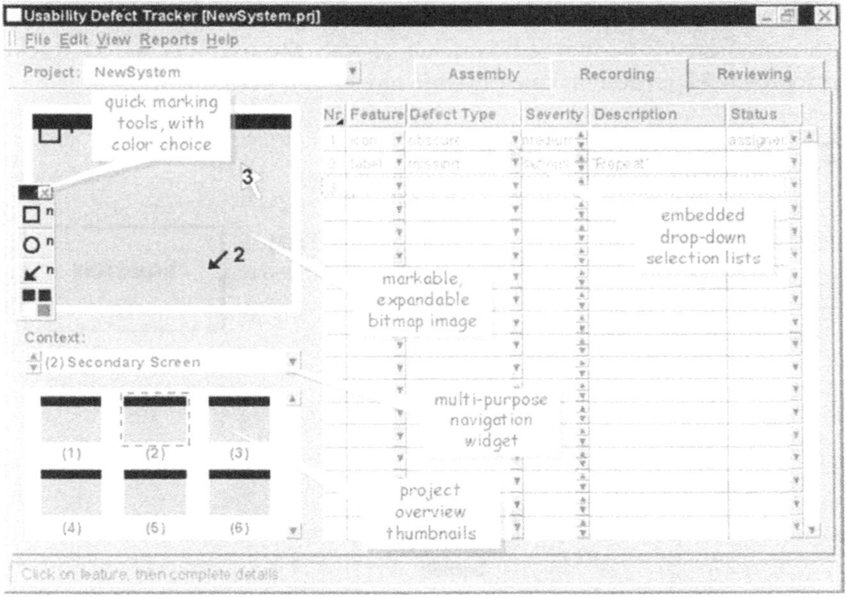

The visual prototype in Figure 4.13, one of many possible designs, incorporates a number
of these "creative translations." The defectLocationMarker becomes a floating tool palette;
the usabilityProblemTypes and usabilityProblemList have been combined using embedded
drop-down lists. Usability inspections and testing could most likely further improve this initial visual design.

Prototype Payoff

Abstract prototyping helps you quickly figure out what you're building before you start positioning and linking GUI widgets with a visual development tool. Based on our experiences,
you should get superior user interfaces in fewer iterations: designs that are simultaneously

more innovative and a better match to user needs. The small investment in constructing a content model can lead to big-time savings over the entire project. Visual programming goes even faster when you already know what screens, windows, and dialog boxes need to be created and what needs to go into each one. Remember, too, all the time you spend building the wrong prototype is wasted anyway!

4.4.5 "Breaking the Barriers"

by Stephen Holmes, August 1998

Reach users around the globe by taking these steps toward creating internationalized software.

As the world becomes smaller, due in large part to corporate globalization, software from multiple locales must interact correctly with existing systems — and each other. When developing or enhancing global applications, you have to deal with multiple date and time formats, currencies (not to mention the Euro dollar), taxation rules, name and address formats, and more. So much thought has to go into the design and planning process that you'll reel with information overload.

In the current Internet era, we've witnessed the birth of online banking, e-trading, e-cash, and a multitude of heterogeneous database systems — all interconnected on the World Wide Web. As a European trying to purchase a book from the World Wide Web, I regularly become frustrated trying to fill in a U.S. postal address requiring a ZIP code. We don't have ZIP codes where I live, so I either make one up or have my form rejected. This is just a tiny example of what we're all up against.

Suddenly, developers are confronted with a new force to reckon with; one with verbiage to match. Localization, internationalization, character sets, and simultaneous shipment jostle for position in your technical vocabulary.

In the rush to get the job done, many companies try to ramp up an internal translation division — with a minimal budget, scarce resources, and crazy schedules — to take the existing product and endeavor to translate it perfectly. Unfortunately, as most localization groups realize, this proves to be fraught with more problems than anyone could have imagined — translators, for one thing, are usually language graduates, not C++ gurus. Another stumbling block is trying to convince development that localization is more than translation. Of course, many companies decide they don't need the hassle and they outsource the entire effort to a partner company.

Picking a Suitable Model

So how do you decide which model best suits your organization? Let's take a look at a couple scenarios.

First, imagine you have a group of developers concentrating on rolling out software internally, say a Year 2000-compliant payroll system, and you need it to work with your corporation's sites worldwide. You might consider contracting an internationalization specialist to work on-site with your team to assist in getting your software globally ready.

Second, if you've already finished your product and just want to ship it internationally, an outsourced model would probably be more appropriate. If your software is domain-specific (medical, accounting, or aerospace, for example), you may need to have one of your team members work closely, perhaps on-site, with the outsourced partner. This can ensure functionality is not broken during the internationalization effort. Typically, you would provide the specification documents and full source build environment. Don't forget to talk to your legal folks about the ramifications arising from this kind of work, however.

This article focuses primarily on the issues related to taking an existing product to a global audience.

Where to Begin

Clients often deliver software to me for translation with a confident claim that it's ready to translate, meaning it's fully internationalized. I usually nod in agreement and take it away for my own analysis. As part of this analysis, I perform a simple experiment using available tools (such as Microsoft RLMAN or Corel Catalyst). To begin, I substitute all visible user interface elements (dialogue titles, static text, button text, and so on) with pseudo-translated strings. These pseudo-translated strings are pre- and post-fixed with a '#' character to make them stand out prominently. You can easily see the changes to the resources when the modified application is launched. The actual strings between the '#' delimiters also have their vowels replaced with suitable extended characters at the same time, for example an 'é' for an 'e.' I take this extra step during pseudo-translation to ensure the system correctly handles strings that contain ANSI values greater than the ASCII 127 limit. I then launch the modified application and start to look for the telltale English strings that are still visible. Figure 4.14 illustrates how you can spot hard-coded strings.

Figure 4.14   Spotting hard-coded strings.

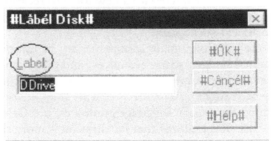

The development group must ensure that no hard-coded strings exist and that everything requiring localization modifications has been separated into resource files. The next part of the experiment involves launching the original application on a foreign language operating system (for example, the Japanese edition of Microsoft Windows NT). In this experiment, I look for the usual international gotchas such as directory naming conventions (for example, on localized versions of Windows 95 and Windows NT, the Programs Files folder is translated), date, time, and currency formatting; and general extended character handling. If you have a native language speaker available on your team, you might want to check the input/output features of the software to ensure it's as functional on the localized platform as it is on the U.S. platform. These input/output tests might involve input method editors for

Asian character input or general foreign language input and output operations such as file names containing non-English characters.

These tests won't yield a perfectly accurate reflection of the product's current level of internationalization, but they will highlight some of the more common downfalls encountered by companies trying to localize their existing software.

This experimentation does not take too long, but can save a lot of heartache once you enter a translation production cycle. Throwing a poorly internationalized product into translation will introduce the following cycle: translate, isolate internationalization issues, fix international bugs, hand-off an update, and translate new strings in update. This can only lead to protracted delays, higher costs, and a frustrated production staff. Here's a tip: Updates during the localization cycle cost money; internationalize first and perform the smoke test discussed earlier to ensure you've caught as much as possible before handing off files to translation.

In a corporate environment, it makes sense to recruit an internationalization specialist or consultant to work with your team. He or she should be responsible for performing these smoke tests during the coding effort. You should subject each new build to a number of compliance tests, some of which are outlined in this article. The specialist should be able to converse with your team on all aspects of your software that will be affected in the effort to globalize. You should include input from the specialist in the project plan, and specify all design decisions regarding the internationalization effort in the project's design documents.

Going from Scratch: Build a Strong, Knowledgeable Team

I cannot emphasize the importance of fully integrating the localization and development teams into the project development life cycle. Let them define their requirements for localization throughout the project plan — not just as an afterthought in the appendix. Have your development staff adhere to some of the basic rules of international coding, and introduce international sufficiency tests into your standard product testing cycles during the entire development effort. This means you'll probably have to get training and institute new quality control procedures for the developers. Focus on the rewards of an international presence rather than the up-front cost. A typical team would include:

- An internationalization development engineer, who oversees and reviews the code from an international perspective
- An internationalization quality assurance engineer, who produces international test data sets and writes or rewrites automated scripts for international applications
- A translation and tools expert, who can assess the actual translatability of the product from a language and tools perspective.

To build a knowledgeable team, code to international standards from the first line and check international functionality regularly.

The Internationalization Approach

Before embarking on any localization effort, you would be wise to take a long and detailed look at your existing code base. Tackling key problems in a structured manner will help you bring your internationalized code base to fruition. To structure your internationalization process, you will need to apply the following steps to your code base, and integrate unit tests that will sufficiently test the code at each stage of development.

With existing bodies of code, I have found that phasing in each of the required code modifications, one step at a time, helps me avoid overcomplicating the overall project. The typical phases in an internationalization process are illustrated in Figure 4.15.

Figure 4.15 Five stages of internationalizing existing applications.

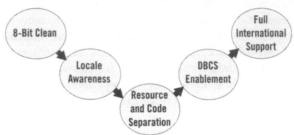

Step One: 8-bit clean. Ensure no assumptions are made on the size of a character. Although I say 8-bit clean, this is really about removing all assumptions about character sizes. Also be aware that you can combine this step in a single update effort with Step 4 (enabling double-byte character sets). Bear in mind, however, it's much easier to retrofit an application to work with single-byte languages than to make a one-fits-all solution in one pass. For example, Microsoft has three versions of Windows 95 (Pan-European, Asian, and Middle Eastern). During the 8-bit cleaning stage, all modules in your code that process or parse strings should be identified and, if possible, separated into a discrete string handling module that all subsequent modules can reference. Modular code is always encouraged in companies that develop software, so continue with that mantra for all international dependencies. Typical string handling code should provide for at least sorting, parsing, comparing, and formatting.

Step Two: Locale awareness. Virtually every modern operating system provides a set of APIs to deal with the notion of locale. Typically, an operating system will hold all locale information in a data set that details such things as the formats of currencies, dates, and times, as well as preferred calendars and code page mapping information. Depending on the operating system in question, the API may provide more or less functionality. Certainly, the offerings from Microsoft have well-developed national language support (NLS) APIs. The NLS API provides functions like `getlocale()`, `setlocale()`, `GetCurrencyFormat()`, `GetDateFormat()`, and so forth. Windows NT 5.0 (whenever it arrives) will take international support to new levels.

Step Three: Resource code separation. Ask anyone at a translation services company and they will tell you about all of the instances of projects that required in-code localization. Separating all user interface elements, messages, and general strings increases the product translatability and moves it to the next level of internationalization. Leaving hard-coded strings causes frustration and last-minute code modifications during the translation cycle — a recipe for disaster.

Step Four: Double byte character set (DBCS) enablement. Although linked very closely with Step 1, DBCS enablement becomes a much easier task when the application is already 8-bit clean and separated from its resource base. DBCS enablement can affect your code base in a number of ways. For instance, strings can no longer make assumptions on buffer sizes or word start and stop boundaries. Even at the individual character level within a DBCS string, you will have to make provisions for special parsing routines. For example, one of the most common operations performed on a string might be something like:

```
while (*p++)
    // do something useful
```

If the string is translated into Korean, this routine will no longer be reliable. This is because a Korean character could be one, two, or more characters long. Remember, in a character set like Korean, there are more than 11,000 characters. The only way to represent them is to string a number of them together that, in sequence, render a single displayable character. You can discern whether a character is displayable by finding out if it forms the start of a sequence. If it does, it is called a lead byte. Similarly, you know that you've reached the end of a sequence when you reach a special character called the trail byte. All string processing functions need to take this into account when the string is being handled. Microsoft, in its C libraries, provides functions such as CharNext and CharPrev. Using these, you can be confident that the string will be processed character by character instead of byte by byte. The correct code would then be as follows:

```
while (CharNext())
    // do something even more useful
```

Step Five: Full international support. A fully internationalized application should be able to run on the localized edition of an operating system and accept characters from that locale (whether they're European single-byte characters or Asian combined characters). Having your application perform this feat means you will need to support the following:

- Input method editing (on Windows, you would process the WM_IME_XXX messages)
- The potential for vertical writing (on Windows, the font could be @MS Mincho, where the @ signifies a requirement for vertical text)
- Multilingual input/output requirements in your application (maybe it is a word processor) may force you to adapt your file formats to include code page information (such as RTF) so people can read all the files, regardless of application, without trouble.

Working with an Existing Product

Let's face it, it's a minority of software companies that design their products with full international support at code-zero (the project concept and development plan). For most, focusing on the U.S. product release is paramount. The pressure for sales comes from the U.S. and English-speaking markets. Unfortunately, there is a common misconception that supporting international functionality is a feature enhancement; thus, it gets dropped at the blink of an eye in favor of Super-Cool Feature X. Of course, while the U.S. product may be a technical marvel, by the time it is localized, you end up running barefoot over a bed of hot ashes. The most common problems existing in code bases today include:

- *Hard-coded strings.* An example is illustrated in Figure 4.14.
- *Dynamic sentence construction.* Many languages use a different verb and noun order. Assumptions made on the ordering will make a correct translation impossible.

- *Fixed-size or dynamically constructed user interface elements.* If you fix the size of a dialog in code, it doesn't allow for a typical string expansion of 20% to 50% (depending on the language). Placing controls into the user interface at run time also causes a sizing issue because they have to be located in space that can accommodate them.
- *Poor support for international conventions.* This is a general way of saying the application ignores facts such as currency issues, dates, times, list separators, and external files.
- *Limited support for foreign characters in strings and string processing functions.* This is one of the most common problems that occur in software. The manner in which a string is handled can have adverse effects on translated materials. For example, the following code fragment attempts to check for a valid letter:

```
if (ch >= 'a' || ch < 'z')
    isValid();
else
    isInValid();
```

In other languages, the alphabet has more characters than just "a through z," and this code won't honor those characters falling outside the standard English range.

International Testing Requirements

Testing in software development is critical. The process and implementation of testing often isn't well understood. How well testing is implemented varies from one company to the next. However, when we are testing an internationalized or localized product, it is vital to take some things into consideration.

First, be aware that automated test scripts, like the code being tested, must be internationalized. Companies involved in software development typically have a body of test script code developed for the U.S. version. Unfortunately, I've found the scripts will not work on localized operating systems or with the localized application. In essence, the same ground rules apply to script code and application code. You will need to separate all the entities that can be localized from your script code into separate files or modules. You will also need to test your test scripts against the localized product.

Next, the test team should have notes detailing all locale-dependent entities. You should document common international attributes, such as date and time formats, currency handling, extended character tables, and input/output requirements in the test plan for each target locale.

Lastly, some applications rely heavily on input/output processing. With software of this type, you will need to collect and validate international test data to run through the system. For instance, you may want to ensure what goes into the database comes out the same way.

Going for Corporate Buy-In

Internationalization is not rocket science. What it boils down to is a conscious decision to develop a global application with the buy-in and support of management and implementation staff. You'll also need a domain expert. Be prepared, though, these people are rare and command a steep hourly rate. If you can develop the experience internally, do that instead.

Obtaining buy-in also requires breaking through your organization's politics. You can jump start this process by measuring the requirements for your international product, involving an internationalization consultant with demonstrated implementation experience, and then assessing the application for its current level of internationalization.

Next, you can outline the benefits of a global roll-out. You can do this by collecting the data you'll need in an international document. This document should become an integral part of the development and testing project plans. Testing is crucial, so get your international data sets ready to run against your application.

Lastly, you must write and present a high-level plan to your organization. Once you've obtained buy-in and developed a detailed roll-out plan, the final step is to get your team's approval on tasks and milestone dates. At that point, you'll be well on your way toward successful internationalization.

Internationalization Testing Checklist

Cosmetic/Visual:

- Check for text that isn't translated.
- Translations should, at a minimum, meet the standards of native speakers with respect to the chosen grammar and the accuracy of the terminology.
- Dialog boxes should be properly resized and dialog text hyphenated according to the rules of the user interface language.
- Translated dialog boxes, status bars, toolbars, and menus fit on the screen at different resolutions. They do not wrap and are not cut off.
- Menu and dialog box accelerators are unique.
- Visual layout is consistent with the native edition's layout. For example, dialog box elements are in the proper tab order.

Functional:

- The user can type accented characters and long strings in documents and dialog boxes.
- The user can enter text, accelerators, and shortcut key combinations using international keyboard layouts.
- Documents created in this language edition can be opened successfully in other language editions, and vice versa.
- The user can enter dates, times, currencies, and numbers using international formats.
- Sorting and case conversion are culturally accurate.
- Paper size, envelope size, and other defaults are culturally accurate.
- The user can save files using names that contain accented characters.
- The user can successfully cut and paste text that contains accented characters to other applications.
- The application responds to changes in the Control Panel's international and locale settings.
- The application works correctly on different types of hardware, particularly those sold in the target market.

- The U.S. application should work correctly on all localized editions of Microsoft Windows and the language edition of Windows that corresponds to the language edition of the localized application itself. The same ground rules apply to other platforms.

General International Conventions:
This list summarizes the main areas you'll need to focus on when internationalizing your application. Although the list is not exhaustive, it provides a base from which you can assess the individual requirements in your own software.

- *String handling.* Make sure you've allocated enough space in your string buffers. If you are using Java or C++, use String classes that can grow to accommodate the translated strings transparently.

- *Date, times, and currencies.* These are the general areas that are well documented in most programming languages or operating system APIs. Research them and figure out how you will standardize their use. With currencies, remember that not every locale uses two decimal places and that the positioning of the currency symbol can change.

- *Separate your resources.* Most modern programming languages support a resource model (.RC files in Windows or .properties files in Java). Place all of your elements that can be localized into these files. When you translate them, you only have to distribute and recompile these elements.

- *Enable input methods.* Although this really affects Asian languages, it is a good practice to implement the input method editor (IME) interface in your code. Both Microsoft Win32 APE and the X11 protocols provide detailed support of IMEs.

- *Be conscientious about your user interface design.* Don't clutter dialogs and don't make ambiguous use of your own language. Develop an English glossary of terms and only add to it with team approval. Also avoid using acronyms.

Chapter 5

Best Practices for the Infrastructure Management Workflow

A significant difference between the enhanced lifecycle for the Unified Process (depicted in Figure 5.1) and the initial lifecycle is that the enhanced version recognizes the fact that most organizations have more than one software project that they need to manage. The reality is that software project teams are dependent on the infrastructure of your organization and on one another from a resource-sharing point of view. At the same time, within their defined-scope software project, teams are also independent of one another and free to make decisions in an appropriate manner. In short, for a software process to be truly effective, it must include activities that support the successful management of portfolios of software projects. The Infrastructure Management workflow was added to the enhanced lifecycle of the Unified Process to focus on the management and support of cross-project issues.

Project teams are both dependent and independent.

The activities of infrastructure management are summarized Table 5.1, and as you can see, there are a wide variety of important activities that this workflow encompasses. Because the Elaboration phase concentrates on the development of the architecture for your project, you will focus your efforts on reuse management and organization/enterprise-level

131

architectural modeling for the Infrastructure Management workflow. Through your reuse management efforts, you will identify existing artifacts, such as common frameworks and components, that you can likely apply to your project efforts. You should also start with your organization/enterprise-level domain and technical architectures as the basis from which you develop the specific architecture for your project as there is no need to reinvent the architectural wheel. Naturally, as your project's architectural efforts proceed during the Analysis and Design workflow (discussed in Chapter 6), you will want to update your organization/enterprise-level architectures appropriately.

Figure 5.1   The enhanced lifecycle for the Unified Process.

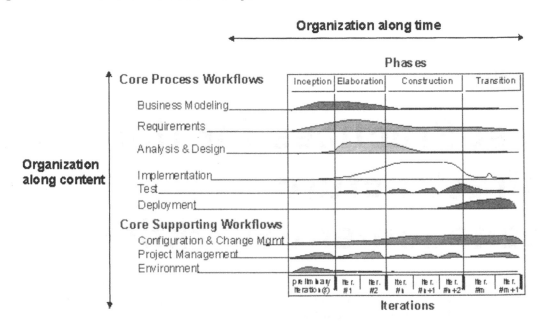

The Infrastructure Management workflow encompasses the activities that typically fall outside the scope of a single project.

Table 5.1   The aspects of infrastructure management.

Activity	Definition
Strategic Reuse Management and Support	The identification, generalization and/or development of, and support of potentially reusable artifacts. This topic is covered in detail in Volume 3: *The Unified Process Construction Phase*.
Software Process Management and Support	The identification, documentation, and support of a software process to meet your organization's unique needs. This is often the responsibility of a Software Engineering Process Group (SEPG). The software process itself may be tailored to meet the specific needs of an individual project as part of the Environment workflow. This topic is covered in detail in Volume 4 in this series, *The Unified Process Transition Phase*.
Enterprise Modeling	The modeling of the requirements for an organization and the identification and documentation of the environment that the organization operates within. The purpose of enterprise modeling is to understand the business of the organization to direct the development of a common, reusable architecture that supports the needs of your organization and to direct your programme management efforts.
Organization/Enterprise-Level Architectural Modeling	The development and support of both a domain architecture that supports the business of your organization and a technical/system architecture that provides the underlying technical infrastructure for your systems. These models enable your Enterprise Application Integration (EAI) efforts, providing guidance for developing new applications based on a common architecture and for integrating legacy applications into overall infrastructure.
Standards and Guidelines Management and Support	The identification, development/purchase, and support of the standards and guidelines to be followed by software professionals. These standards and guidelines may be tailored to meet the specific needs of a project team as part of the Environment workflow.
Programme Management	The management of the portfolio of software of an organization, including legacy software that exists in production, software projects currently in development, and proposed software projects awaiting development.

5.1 Enterprise Requirements and Architectural Modeling

Domain component reuse is a highly effective form of reuse — a type of reuse that is obtained as the result of organization/enterprise-wide (enterprise) architectural modeling. A large-scale component is a collection of classes that together provide cohesive behavior. A bank may have domain components such as Customer, Account, and Portfolio Management whereas a manufacturing company may have domain components such as Customer, Supplier, Inventory, and Shipping. Both organizations may have similar technical components/frameworks such as Security, System Manager, and Transaction Manager. The development of large-scale, common, reusable components and frameworks require effective infrastructure management to achieve — reusable domain components and frameworks only pay for themselves when they are used by several projects. To effectively identify large-scale components, you must first understand the requirements that you are trying to support, and because large-scale components are meant to support your organization as a whole, your enterprise-level requirements should be your starting point.

Infrastructure-driven modeling is a key enabler of high-impact reuse.

In the articles "Object-Oriented Architectural Modeling: Parts 1 & 2" (*Software Development*, September 1999 and October 1999), I present the fundamental concepts of infrastructure-driven modeling. These articles present a detailed look at the processes and organization structures needed to support infrastructure-driven modeling, as depicted in Figure 5.2. The reality is that architecture is both an organization/enterprise-level infrastructure issue and a project-level modeling issue — organizations that do not recognize this fact will never be able to take full advantage of the benefits of an architecture-driven approach. Part 1 lists the potential artifacts of the various modeling processes — a valuable resource for organizations that are evolving their software processes to reflect the needs of their component-based and object-oriented software infrastructures. Part 2 discusses in detail the cultural issues that your organization will likely face with respect to the Infrastructure Management workflow; issues such as educating senior management and your development staff as well as how to organize your project teams effectively. Modeling tool support is also discussed in the second article.

So how will this process work in practice? Let's go through an example. In Figure 5.3 a potential approach to the infrastructure-driven development process is presented. For the sake of simplicity, the technical architecture aspects are left out of the picture because they effectively take on an equivalent role as domain components within the process. Taking a top-down approach to development, you would start with the development of an enterprise requirements model — the primary artifacts being an enterprise use case model and an enterprise business process model. The enterprise model would then be analyzed to identify candidate domain classes — potentially being used to form an enterprise domain model, which, in turn, would be grouped into a collection of domain components modeled by your business/domain architecture model. Once this effort is initially completed, your organization will be in a position where it understands both the enterprise-level requirements describing your business environment and the "shape of the landscape" of the business-domain components that will be implemented to support those requirements. You also need to perform similar work for the technical side of things — modeling enterprise-level technical requirements and the technical architecture "landscape".

Figure 5.2   Infrastructure-driven modeling process.

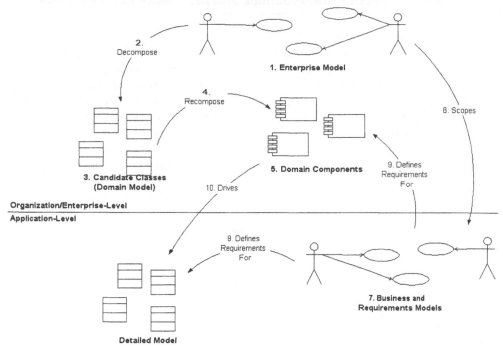

Once these efforts are completed, you will now have an understanding of the overall domain and technical infrastructure needed to support your project teams. The project teams would then come along and start by scoping out a portion of the enterprise model — a goal of the Requirements workflow during the Inception phase, typically small slices of the behavior of several enterprise-level use cases. The project team proceeds to flesh out the Business Model and the Requirements Model for their application, potentially producing a detailed use case model describing the business functionality their application will provide to their users. This requirements model is then analyzed to determine *a*) which requirements are currently fulfilled by existing functionality in the domain components, *b*) which functionality is currently allocated to domain components but has not been implemented, and *c*) which functionality has not been previously identified. Existing functionality should obviously be reused and identified but unallocated functionality needs to be scheduled for development (work which would be developed by one or both of the project team and/or the infrastructure group). The previously unidentified functionality falls into one of three categories: behavior that belongs at the enterprise level to be shared by several applications, behavior that belongs within a single business unit that is applicable to several applications in that business unit, or behavior that belongs within a single business and that is specific to a single application. This behavior then needs to be scheduled to be added to either an enterprise-level domain component, a business-unit specific domain component, or to the specific application code respectively. Categorizing and scheduling this work is naturally a change-control activity of the Configuration and Change Management workflow. Development work would then proceed appropriately.

Following common practices and processes enables reuse.

In section 5.3.3 "A Hybrid Approach" (*Software Development*, September 1998), Lawrence Finch presents a component-based architecture consisting of vertical components and horizontal frameworks. A vertical component implements cohesive business functionality, similar in concept to business/domain components. Horizontal frameworks implement your technical architecture, including features such as your persistence layer and security facility. In a sidebar, he puts horizontal and vertical frameworks into the context of Enterprise JavaBeans (EJB). Finch describes the importance of the best practice of developing technical specifications for your horizontal frameworks — you need to take a requirements-driven approach to be successful. This is a concept that is near and dear to my heart, as I have seen several technical architecture efforts go awry due to a lack of requirements. The bottom line is that if your design isn't based on documented, accepted, and well-understood requirements, then you are hacking, even if your job title is "Architect." Architecture without requirements is something that Dave Moore, a friend of mine, calls "hacking in the large." Finch also points out that architecture teams can go astray when they only focus on horizontal/technical frameworks, e.g., having a lot of fun playing with the technology, but ignoring the vertical/domain components. With respect to this issue, the bottom line is: the vertical/domain functionality is what offers the real benefit to your organization; the horizontal/technical frameworks are merely enablers. In short, this article effectively brings together many concepts of enterprise-level architectural modeling and shows how components and frameworks can be used together to achieve this goal.

Architecture that isn't based on requirements is simply hacking in the large.

How do you guarantee that your architecture is in fact robust and extensible? This issue is directly addressed by development of change cases, a topic that I discuss in section 5.3.4 "Architecting for Change" (*Software Development*, May 1999). Change cases, presented in detail in the book *Designing Hard Software* (Bennett, 1997), are a technique for documenting likely changes that your software will one day need to support. A change case model is effectively a modification of a use case model that enables you to document the results of the brainstorming that architects and designers typically perform regarding potential changes that your system will need to support — allowing you to take these important considerations into account when designing your software. Change cases can be developed for both domain issues and for technical issues. For example, the domain architects at a bank may wish to consider the question of whether the bank is likely to sell insurance to its customers or to allow their customers to work with any type of currency in any type of bank account. The technical architects at the same bank may wish to consider the question of whether they need to support Internet-based access to their systems or the need to support distributed processing. Each change case, often documented as a single paragraph, provides the basis from which architects can judge whether or not their models are robust. Change cases also provide senior management the information they need to understand why certain design decisions were made, enabling them to judge whether or not the decisions were reasonable or "gold-plating" (the addition of unnecessary features) on the part of the architects.

Change cases enable you to judge the soundness of your architectural models.

5.2 Integrating With Legacy Applications: EAI

Few organizations have the benefit of starting from scratch, a concept called "green-field development," but instead must integrate their new work with existing legacy applications. In section 5.3.5 "Enterprise Application Integration From The Ground Up" (*Software Development*, April 1999), the first of a three-part series, David S. Linthicum presents an overview of the issues surrounding enterprise application integration (EAI). The purpose of EAI is to build seamless bridges between applications within your organization — to integrate your disparate applications in such a way as to make them appear to be one, an effort that obviously spans several projects. As organizations move into the hyper-competitive world of the Internet age, they are discovering that their stand-alone "stovepipe" applications (those that address one specific business need that do not interact with other applications) are insufficient for their modern-day needs. In this article, Linthicum describes EAI as the "unrestricted data sharing among any connected applications or data sources in the enterprise. You need to share this data without making sweeping changes to the applications or data structures. In other words, you must leave it where it lies." This is a tall order, and to his credit, Linthicum states that despite the marketing claims of technology vendors, there is still no solution to the EAI problem. He does however lay the groundwork for how to go about EAI, presenting a layered approach for organizing your efforts. The majority of the article focuses on: middleware issues, overviewing common communication techniques such as remote procedure calls (RPCs), message-oriented middleware (MOM), transaction processing, database-oriented middleware, message brokers, and distributed object techniques such as Microsoft's DCOM and the Object Management Group's CORBA.

EAI is the unrestricted data sharing among any connected applications or data sources in the enterprise without making sweeping changes to the applications or data structures.

Linthicum continues his three part series in section 5.3.6 "Mastering Message Brokers" (*Software Development*, June 1999) by showing how to apply message brokers — pieces of software that act as brokers between target and source entities (e.g., between user interface components and large-scale business/domain components) to share information effectively. Linthicum discusses the architecture of message brokers, describing message transformation, intelligent routing, rules processing, message warehousing, repository services, directory services, and message broker adaptors. He also overviews potential configurations for deploying message brokers, such as a bus or hub and spoke approach — important information for anyone defining their organization's EAI architectural strategy.

Finally, in section 5.3.7 "Building Enterprise Architecture" (*Software Development*, September 1999), Linthicum finishes his series by addressing key, fundamental architectural questions that many organizations face today. A significant best practice presented in the article is the need to think about architecture first — a concept that is fundamental to the enhanced lifecycle for the Unified Process. Linthicum covers process and methodology issues pertaining to EAI, comparing and contrasting approaches such as data-level EAI, API-level EAI, Method-level EAI, and User Interface-level EAI. There are several strategies for approaching EAI, and you need to understand their strengths and weaknesses to choose the approach that is best suited for your organization. The article finishes with a discussion of the outstanding issues involved with EAI, such as the need for industry-standard approaches to technical architecture ("plumbing"), process automation and workflow, and industry-standard data sharing formats such as extensible markup language (XML).

There are four major approaches to EAI, each with strengths and weaknesses.

I would like to end this chapter introduction describing a fundamental best practice to ensure the success of your organization/enterprise-level architecture efforts: they must be based on requirements. Many architects fall into the trap of not basing their work on requirements, resulting in work that either doesn't meet the needs of application teams or that appears to be overkill for their needs. Many enterprise domain architecture efforts went astray in the 1980s and 1990s because of this issue. The main artifact produced by these efforts was often an "enterprise data model" — a model that purported to model the data required by the organization. Although these models were often very good, they unfortunately were rarely traceable to the business requirements that they were trying to support (even though the modelers may have implicitly understood these requirements, they often didn't think to explicitly record them). The end result was that the "enterprise modelers" didn't have a story to tell. So when a project team came along, they couldn't easily map their project's requirements to the enterprise model, allowing the project team to claim that the enterprise model didn't truly meet their needs or constrained them in some manner so, therefore, they shouldn't conform to it. The end result was that the enterprise modeling efforts failed. Technical architects that put together a sophisticated technical architecture based on their experience and knowledge make the exact same mistake. Although they often do an excellent job, once again, they have based their work on their implicit understanding of the technical requirements instead of starting their effort by defining them explicitly.

Your models should be based on requirements, even architectural models. Without defined and accepted requirements, you are "hacking in the large".

5.3 The Articles

5.3.1 "Object-Oriented Architectural Modeling: Part 1"

by Scott W. Ambler, September 1999

There are 10 ways you can model your development projects.

These days it seems like everything is "architecture-centric" or "architecture-driven," but what does that really mean? Why is architecture important to you? How does architectural modeling fit into your overall development process? It's likely that your organization has been struggling with these questions for years, and I hope I can shed some light on how architectural modeling works in the real world. Although this column is targeted at organizations using object or component technology to develop systems, the material is high-level enough to work in any environment — your development deliverables may change, but the fundamentals remain the same.

I'll start with the basics of the modeling process, captured in the Model process pattern depicted in Figure 5.3. It is based on David Taylor's "T" technique, in which enterprise and organizational modeling efforts, forming the top bar of the "T," drive detailed and project modeling efforts, each of which form the trunk of the "T." This pattern shows that there are several different styles of modeling that work in both a top-down and a bottom-up fashion. The Model process pattern adds greater detail to Taylor's original approach.

Figure 5.3 shows what I consider the 10 styles of modeling — enterprise modeling, four types of architectural modeling, four types of detailed modeling, and requirements modeling. It also shows that there are two fundamental levels of modeling, organizational level and project level. Let's explore each level of modeling in detail. Table 5.2 summarizes the inputs and outputs of each modeling style — although many of the artifacts are similar, their level of detail and how they are used change between styles.

Figure 5.3   The Model process pattern.

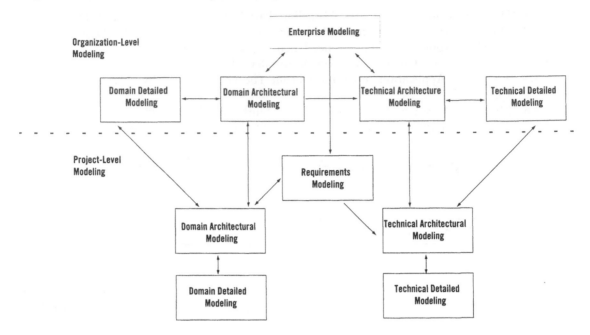

Table 5.2   The artifacts of object-oriented modeling.

Input	Process	Output
•Existing "Enterprise Models" •Glossary	Enterprise Modeling	•Enterprise Model - Use-Case Model - Business Process Model - Change Case Model •Glossary
•Enterprise Model •Glossary	Organization-Level Domain Architectural Modeling	•Domain Architectural Model - Component Model - Sequence Diagram(s) •Glossary
•Domain Architectural Model •Enterprise Model •Glossary	Organization-Level Technical Architectural Modeling	•Technical Architectural Model - Component Model - Infrastructure Model - Deployment Model(s) - Framework Model(s) - Sequence Diagram(s) •Glossary

Input	Process	Output
•Organization-Level Domain Architecture Model •Requirements Model •Glossary	Organization-Level Domain Detailed Modeling	•Class Model(s) •Sequence Diagrams •Collaboration Diagram(s) •Statechart Diagram(s) •Activity Diagram(s) •Physical Persistence Model •Glossary
•Organization-Level Technical Architectural Model •Requirements Model •Glossary	Organization-Level Technical Detailed Modeling	•Class Model(s) •Sequence Diagrams •Collaboration Diagram(s) •Statechart Diagram(s) •Activity Diagram(s) •Framework Model •Glossary
•Domain Architectural Model •Requirements Model •Glossary	Project-Level Domain Architecture Modeling	•Project-Level Domain Architecture Model - Component Model - Sequence Diagram(s) - Change Proposal(s) •Change-Case Model •Glossary
•Technical Architecture Model •Requirements Model •Glossary	Project-Level Technical Architecture Modeling	•Project-Level Technical Architecture Model - Component Model - Deployment Model - Sequence Diagram(s) - Change-Case Proposal(s) •Glossary
•Enterprise Model	Project-Level Requirements Modeling	•Use Case Model •Supplementary Specifications •User Interface Mockups •Business Process Model •Business Rule Dictionary •Glossary

Input	Process	Output
•Project-Level Domain Architecture Model •Requirements Model •Glossary	Project-Level Domain Detailed Modeling	•Class Model(s) •Sequence Diagrams •Collaboration Diagram(s) •Statechart Diagram(s) •Activity Diagram(s) •Physical Persistence Model •User Interface Model •Glossary
•Project-Level Technical Architecture Model •Requirements Model •Glossary	Project-Level Technical Detailed Modeling	•Class Model(s) •Sequence Diagrams •Collaboration Diagram(s) •Statechart Diagram(s) •Activity Diagram(s) •Framework Model •Glossary

Organizational-Level Modeling

The scope of organizational-level modeling is that of your entire company, the goals being to model your fundamental business with an enterprise model, the software that supports your business in a domain architectural model, and your system infrastructure in a technical architecture model. As shown in Figure 5.4, there are five styles of organizational-level modeling: enterprise modeling, domain architectural modeling, domain-detailed modeling, technical architectural modeling, and technical-detailed modeling.

The goal of enterprise modeling is to model your organization's high-level requirements so you understand the fundamentals of your organization and its environment. To do this, you must develop three artifacts: an enterprise-level use case model, an enterprise-level business process model, and an enterprise-level change case model. The use case model will depict the high-level business behaviors that your organization supports; for example, a bank would likely have use cases such as "Manage Money," "Manage Loans," and "Market Services." Your business process model models the business processes that support the required business behaviors and the business environment that your organization operates in. For example, a bank's business process model would show processes such as "Process Loan Payments" and "Bill for Loans," depicting the information flowing into and out of those processes. The difference between the two artifacts is that your use case model focuses on what your organization does, whereas your process model focuses on how you do it — in other words, one focuses on requirements and the other on analysis. The third artifact is the change case model (see section 5.3.4 "Architecting for Change" on page 157), which documents likely changes to your business and technical environment, information you need to develop a robust architecture for your organization. It lets you develop models that easily support these potential changes.

Figure 5.4   Object-oriented architectural modeling overview.

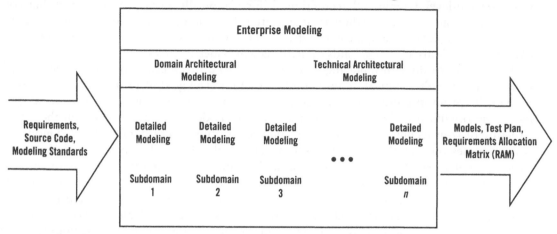

In previous writings, I have said that a use case model is the primary artifact of enterprise modeling and have generally ignored business process modeling. I realize now that this is inconsistent with other advice that I've been given, namely that a single model isn't sufficient for any one style of modeling because you need several views to accurately model something. No matter how thorough you are, no matter how comprehensive a model is, you simply can't depict all of the information you need in one model. A use case model depicts a system's behavioral requirements, a class model depicts the static organization of software, and a data model depicts how data will be stored in a relational database. You need three of these models at some point in the modeling process. In my series of columns about persistence, I claimed that enterprise-level data modeling is fundamentally flawed because it leads to software that doesn't reflect the actual needs of its users (remember, it models data but not behavioral requirements). But the issue is more generic than this: you need several views and models to depict an aspect of your system. Table 5.2 shows that each of the 10 modeling processes produces several primary artifacts.

Domain modeling models how you intend to build the domain aspects of your organization's software. The goal is to define a common domain software environment that all of your applications share, promoting consistency and reuse between applications. You achieve this through two processes: the first being domain architectural modeling that models your software's static structure in the form of a large-scale component model, and sequence diagrams that model the dynamic interactions of components. The component model depicts the large-scale domain components — a bank would have components such as "Account," "Customer," and "Transaction" — that your applications collaborate with to support your business. The sequence diagrams map the behavioral logic of your use cases and the process logic of your business processes, at the enterprise level to the domain components in your domain architecture model. You can find information about this process in my book *Process Patterns* (Cambridge University Press, 1998), Ivar Jacobson's *Software Reuse* (Addison-Wesley, 1997), and *Component-Based Development for Enterprise Systems* (Cambridge University Press, 1998) by Paul Allen and Stuart Frost.

The second process is domain detailed modeling, which, as the name implies, focuses on the detailed modeling of the domain components. Architecture modeling effort focuses on the components' interface; the detailed modeling effort focuses on the components' interior. Organization-level domain architectural modeling is typically a cross-project effort that your architecture team performs, whereas organization-level domain detailed modeling generally supports a project's development efforts. In the second part of this series (beginning on page 145), I will discuss these kinds of management issues in greater detail.

In addition to domain modeling, you also need to do technical modeling, which models your organization's technical or system infrastructure. The goal of technical modeling is to, once again, promote consistency and reuse between applications. There are also two styles of modeling: organization-level technical architectural modeling, which focuses on the overall structure of your technical infrastructure, and organization-level technical detailed modeling, which focuses on how the components and frameworks that make up your technical infrastructure will work. Many developers consider technical modeling the fun part of software development, because you get to define the plumbing of your software, although many make the mistake of ignoring domain architectural modeling. Your technical architecture will often describe your software security approach, how your system will be deployed, your persistence layer or framework, and your system audit component, to name a few. Because your technical architecture addresses a wide range of issues, you need several views. Use component models to show your organization's technical components (such as your security and audit components), framework models to describe common frameworks (such as your persistence and user interface frameworks), and an infrastructure model to tie together the frameworks and components, typically using a layered approach.

Project-Level Modeling

You also need to understand the process of project-level modeling, which starts with project-level requirements modeling. Your requirements effort should flesh out a small portion of your enterprise model, likely working portions of one or more enterprise-level use cases as the scope of a project. This effort, in turn, drives your project-level domain architectural modeling and project-level technical modeling efforts. The goal of these two efforts is to understand how your new system fits into the existing domain and technical architectures of your organization, as well as to determine how your project will change to the existing architectures. As shown in Figure 5.4, your project-level architectural modeling efforts will often drive organization-level architectural and detailed modeling. Your architecture group will frequently need to make changes to its models to reflect the new needs of your project team.

Once you understand your application's architecture, you perform project-level domain detailed modeling and project-level technical detailed modeling as needed. This is similar in concept to your organization-level detailed modeling efforts, except for a greater emphasis on the user interface model, and the fact that it models only a single project and not the entire organization.

Architecting for Success

Before concluding, I would like to make two very important points. First, as I noted earlier, the general processes and concepts remain the same, regardless of your implementation technology. If you are using structured technologies such as COBOL and relational databases or even active objects, you still need to perform these various styles of organization-level and project-level modeling. For structured technologies, you are likely to replace class models, sequence diagrams, and collaboration diagrams with structure charts, while for active objects you are likely to use collaboration diagrams in place of component models. The artifacts may change but the fundamental process remains the same. Second, Table 5.2 shows that a glossary is a key input and output of all modeling processes. By having a shared glossary, you increase the consistency between your development efforts and reduce the overall effort by reusing this artifact.

Architecture is critical to software development success, and your software process must reflect this. There are two flavors of architectural modeling, domain and technical, as well as two levels, organizational and project. Your architectural modeling efforts fit into a larger picture that includes both enterprise modeling and several styles of detailed modeling. Because each type of model only showed a small aspect of your overall system, you need several models to successfully describe your software. In short, architectural modeling is a complex endeavor that your organization must strive to understand. Next month, I will expand on the concepts presented here and discuss the management, cultural, and tool-based issues surrounding architectural modeling.

Recommended Reading

- *Business Engineering With Object Technology* by David Taylor (John Wiley and Sons, 1995)
- *Designing Hard Software: The Essential Tasks* by Douglas Bennett (Manning Publications Co., 1997)
- *The Rational Unified Process: An Introduction* by Philippe Kruchten, (Addison-Wesley Longman 1999)
- *Software Architecture in Practice* by Len Bass, Paul Clements, and Rick Kazman (Addison Wesley, 1998)
- *Building Object Applications That Work: Your Step-By-Step Handbook for Developing Robust Systems with Object Technology* by Scott W. Ambler, (Cambridge University Press, 1998)

5.3.2 "Object-Oriented Architectural Modeling: Part 2"
by Scott W. Ambler, October 1999

Your company's software process should be reflected in every level of its structure and culture.

Terms such as "component-based architecture" and "distributed architecture" have sold senior management more than once on new technology, but architecture is more than just a marketing buzzword. Software's architecture describes the elements (objects, components, frameworks) from which the software is built, the interactions among those elements, the patterns that guide their composition, and the constraints on those patterns. Knowing what architecture is, however, isn't enough. You must also understand the process to define and evolve your architecture, organize your architecture team, build an architectural culture within your organization, and choose appropriate tools to support your architectural efforts.

Figure 5.5 shows the symbiotic relationship between architecture, process, tools, and organizational culture. Any change to one element will eventually result in changes to another.

Figure 5.5   Process, culture, architecture, and tools.

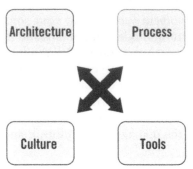

Figure 5.6   The architectural modeling process.

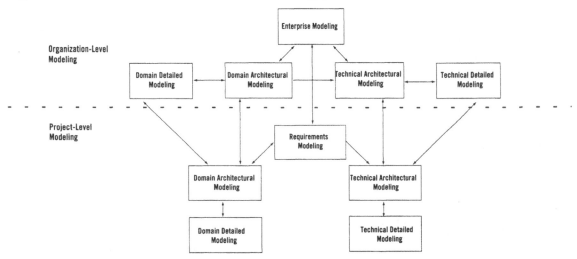

The previous section, "Object-Oriented Architectural Modeling: Part 1," described the architectural modeling process, as depicted in Figure 5.6, and several key concepts of architectural modeling. First, you need both an organization-level and a project-level architecture. Second, there are two flavors of architecture, domain and technical, at each level. Third, your enterprise model defines your organization's high-level requirements and should drive the development of your organization-level architecture. Fourth, your organization-level architecture and project requirements are the starting point for your project-level architecture. And, fifth, your project-level architecture drives change to your organization-level architecture.

What do you do if senior management doesn't see things this way? Managers commonly don't feel the need to invest in architecture, either because they don't have the resources or they believe technology changes too quickly to make the investment worthwhile. I believe this is short-sighted. If you have the resources to build software "the wrong way" without architecture, then surely you have the resources to do it the right way.

Do changing technologies really matter? One thing that the Year 2000 crisis has shown is that the systems you build today will be in production 20, 30, maybe even 40 years from now. It doesn't matter whether the technology you use today will change tomorrow. You'll still be stuck with it — and if you don't use that technology the best way possible, you will have another maintenance nightmare on your hands.

A second common mistake is investing only in project-level architecture, which quickly leads to "stovepipe" systems that don't integrate easily and achieve little, if any, reuse. Project-level architecture is a good start, but not a complete solution. To resolve this problem, you must identify the commonality between systems, and start with technical architecture. It is easy to justify developing or purchasing a common approach to security access control, a shared persistence layer, or a common user interface framework. When management sees many similarities between the applications being developed within your organization, it's easy to convince them to invest in organization-wide architectural efforts. Although you may start with technical architecture, never forget that you also need a domain architecture. You should be prepared to point out similarities in your domain once you have had success on the technical side.

Another common mistake is investing in organization-level architecture but not project-level architecture. Typically, these project teams don't take advantage of their organizations' architecture teams, due to excessive deadline pressure. The architecture team members are often frustrated because their good work is not being used. This problem often stems from an "ivory tower" approach to architecture, in which the brightest and best are sent off to develop an architecture and then asked to bring it back to the programming masses on the project teams. This may be interesting for the people in the ivory tower, but it's not very palatable for the masses.

To address this problem, you must understand what role architects should play within your organization and then ensure that they actually fulfill that role. Architects help define and guide your organizational and project architectures, acting as leaders and key decision makers at some times and as mentors or consultants to your project team at other times. Ideally, architects are responsible both to your organization's senior management and to the analysts and designers of each project team. Architects should be actively involved in establishing your business and technical strategy, as well as your organization's technical direction. The most effective organizations include software experts whenever they negotiate a major contract or define a major organizational strategy. Architects guide technology and architectural decisions, and provide key input into your organization's development guidelines, policies, and processes. In short, architects should be key players within your organization.

When building your organization's architecture team, you must first find good people to be on it. A common joke in the software industry is that you merely need to ask a group of designers to cross a puddle, and then take those who don't get their feet wet because they are the ones who can walk on water. In case this approach doesn't work for you, the best architects are people who not only understand the details of how to build something but know where and when to dive into those details. Architects are abstract thinkers with good technical and communication skills, the humility to understand that they don't know everything, and the confidence to make important decisions when they are needed most. Architects can mentor the people that they work with and be mentored themselves.

It isn't enough to have good people on your team; you must also organize them effectively. Architecture teams should be small, ideally four to six people, and must have a single leader. This person typically has the title of "lead architect" or "chief architect" and is responsible to your senior management for developing and supporting the architecture. Many architecture teams flounder when they try to take a democratic approach to setting architecture — a problem also known as architecture by committee — quickly devolving into endless arguments or producing bloated architectures that contain far more features than necessary.

Your organization's architects will act as consultants to your project teams, either working with them as a full-time member or as a part-time resource. When architects work with a project team full time, they quickly lose touch with the other architects within your organization, falling behind on the latest developments. It also makes it difficult to share the project information that they learn with the rest of the architecture team, which in turn, makes it difficult to get your organization-level architecture updated appropriately. The one advantage to working full time on a project is that it helps the project move forward.

For this architectural approach to work, your organization must accept that projects should be architecture-driven. Developers must also accept that their job isn't to reinvent the wheel, even when they think they can make it better, but rather use the wheels that the architecture team supplies to build a new automobile. Just as architects need humility, so do your project's development staff. As always, the willingness to work toward a common goal is an important cultural value for your organization.

Project teams must develop both a project-level domain architecture model and a technical architecture model, showing how your software uses the organization-level architectures and indicating any proposed changes. You need at least one person in the role of project architect — and, depending on the size of your project, you may need several people. Project architects work with your organization's architects (in small companies this may be the same person) to both exploit the existing architecture and develop a new one for your project. The project architect then acts as a consultant to the developers on your team, sharing and evolving the architecture with them as needed.

To clarify Figure 5.6, part of architectural modeling is analyzing existing systems to understand both the business environment that those systems support and the technical environment in which your system must operate. Ideally, this effort should be done at an organizational level. One aspect of your technical architecture should be a "legacy deployment model" that shows the existing systems within your organization. If you do not have such a model (few organizations do), then your project team needs to identify the legacy systems and their interfaces. These interfaces may not even exist, in which case your team may be given the access rights to directly read from and update the data sources of the legacy systems (yikes!). More likely, there will be interfaces in the form of stored procedures, an application programming interface, or even a full-fledged object-oriented or component interface that you will need to understand. Beware of strangers bearing gifts! I have seen several architecture teams led astray by their data/system analysis efforts — don't let these interfaces overly influence your architecture. Do not let poor design decisions made years ago propagate into your new software.

Modeling Tools

To be effective at architectural modeling, a modeling tool must support both organization-level and project-level models with full traceability between them; support a wide range of models, such as the UML, persistence models, business-process models; and support version control and configuration management of your models. Architectural modeling is challenging for tool vendors. Typically, tools are either targeted at project-level modeling or enterprise architectural modeling, but not both at once. Few tools support multi-team development; in my experience, checking models in and out of a repository simply doesn't scale, nor do many tools support full traceability (or any traceability, for that matter).

Architecture is a complex endeavor, but it's key to successful software development. You need a proven and mature process for architectural modeling, an organization structure that reflects that process, a culture willing to take an architectural approach, and tools that support the goals that you are trying to accomplish.

5.3.3 "A Hybrid Approach"

by Lawrence Finch, September 1998

Increase your team's productivity with a hybrid approach to CBD using horizontal frameworks and vertical components.

The buzzword for 1998 is "components," due in large measure to the potentially positive impact component-based development can have on reusability and development productivity. It wasn't long ago, however, that the same claims were made for object-oriented technologies. Reuse has been the watchword of object-oriented design and development since its inception. However, industry watchers report there's only 15% average reuse in today's object-based projects. A damning statistic considering that, if it's true, developers did better 20 years ago with COBOL subroutines.

Before jumping headfirst into component technology, software teams must plan a course that lets them exploit the potential of component-based development. You can begin by keeping the challenges in mind: software development is more difficult due to increasing requirements and more complex systems. On the whole, system complexity has increased faster than methodologies and tools have improved. With the advent of web-enabled transactional systems, complexity has increased exponentially. Web-deployed applications must be massively scalable, highly secure, and deliverable in Internet years of three to six months. Meeting these requirements forces a move from two-tier client/server architectures to three-tier and multitier architectures. Instead of just a client and a server, there is a browser, a web server, an application server, database servers, and messaging middleware. Today's components must handle the complexity if future benefits are to be achieved from this promising technology.

Components are a step beyond objects. A component is a reusable software building block; a pre-built piece of encapsulated application code that you can combine with other components to produce an application. This sounds like the definition of an object: what differentiates the two is relative scale and how well the implementation is hidden. Implementation hiding is key; a component should be defined well enough so the developer using it does not need its source code. More important, the environment should be able to query the component to determine its needs and capabilities. From this perspective, you can think of components as "big" objects that perform business functions (as opposed to programming functions like controls).

However, components are not as well-defined as objects. There is a well-established theoretical basis for object-oriented methodology — even if some developers don't understand it, don't use it correctly, or disagree with it, there is a body of reference material that precisely defines objects and regulates their use. Components have no such pedigree. (Some would argue that components are what objects were intended to be when they were invented 20 years ago, and that C++ and other implementations of object technology have lost sight of that original intent.)

First, "component" is not another name for "widget" (one common misuse of the word). Components must encompass business functionality. They are made up of business rules, application functionality, data, or resources that are encapsulated to allow reuse in multiple applications. Further, components should be portable and interoperable across applications.

Developers working with components tend to think of them vertically. Consider a bank deposit transaction or a human resources application that completes each task from the user interface to the database update. These components might be implemented as self-contained entities that include the code for all the layers in a multitier environment (the user interface, web server interface, middleware access, and database access). In other words, a vertical component can be defined as an object that implements a business process. It is what the analysis phase of a project tells us we need for a specific business task.

You can define a vertical component in terms of its inputs, business rules, and outputs. It is what an engineer would call a black box — a device that performs a specific function and does not require a look under the hood. Only what it does, not how it does it, matters. When properly designed, vertical components can be inserted into any system and they will do the required processing. Better still, if the business rules change, you can modify the vertical component without affecting any other part of the system.

Any large system consists of some vertical components. Note that when specifying a vertical component, there is little concern over data access, human interface, or the deployment of the application on multiple servers. These are all hidden details that I'll revisit later in this article.

The most commonly used approach (if components are used at all) is to develop vertical components only. But in this case, developers must understand all middleware levels and then code to all middleware and database interfaces. This requires more training and a lot of repetitive code. Many information technology departments are developing software this way because they develop based on functional requirements (which generally define the business aspects of a transaction) rather than on a technical specification.

Technical specifications typically include functional decomposition that lets you develop the application horizontally. Functional decomposition is when you take functional requirements and subdivide them into successive steps: first into business rules, then into specific functional modules, and finally into programming specifications for all system modules.

Most methodologists use this process but call it something else. Object-oriented methodologists may call it use-case analysis, for example. Regardless of name, the critical element is a complete parsing of requirements down to the level of codable modules. Functional decomposition is essential when you're working in a pure object-oriented environment. Many object-oriented projects fail to deliver on reuse because this detailed analysis is bypassed. Without it, commonality of function is not identified, and you lose many opportunities for reuse. But the time to prepare a technical specification is a luxury that many shops cannot (or believe they cannot) afford.

When shops prepare technical specifications, they can partition the work horizontally, with different teams coding each level of the multitier architecture. For example, when developing a multitier banking system, a database programmer might write a service that validates an account number and another that credits a deposit to the database. A GUI programmer will create a screen that collects the account number and amount from the user and forwards that information (using middleware) to a service that contains the business rules for deposits. This service will then call the two database services, get the results, and return them to the GUI piece of the application. If it's a web-based application, a web programmer will translate the GUI data to HTML (or Java) to send to the user's browser. Thus, for one transaction, there are potentially four (or more) programmers and five interfaces involved. Some developers design the database services, design the human interface, manage the web server, or manage the middleware. In other words, the development team structure is horizontal rather than vertical, and so is the end result.

The difficulty with horizontal development is that it shifts the focus from the business problem to the technical details, frequently causing the business need to suffer. Programmers tend to think of their own piece rather than the overall business objective. They worry about messaging middleware issues such as sockets, RPCs, ORBs, or asynchronous messaging and queuing technology. All of these distract the programmer from the real job, which is focusing on business functionality.

In my view, a hybrid approach consisting of horizontal components that encapsulate infrastructure level, non-application specific functions, and vertical components that contain only application-specific functions is better. For this to work efficiently and gain maximum benefit, the horizontal components must communicate transparently with each other and the vertical components.

The concept of vertical components, while compelling, creates its own set of development challenges. At first glance, we have replaced one set of problems with another. True, all the business rules are in one object, but as Figure 5.7 demonstrates, that object has seven or more APIs within it. There is user interface code that uses at least one web API. There is the interface between the web server and the application using CGI, ISAPI, or some other "not quite" standard. There is a middleware client API and a middleware server API. If a mainframe is involved, there is a legacy system API. There is also the relational database API. Finally, there is the API to whatever development tool you are using. Apparently, application programmers must be familiar with all of these APIs (not to mention recovering from errors in all seven levels). Virtually all development tools that use code generation implement vertical components.

When you implement multiple vertical components, each uses the same seven APIs. Also, each must have code to deal with every API, as shown in Figure 5.8. If at some point an API must be altered, you need to correct it in all vertical components. Presented in this context, the vertical component concept by itself is not going to win any awards for efficiency. There is a solution, however, that combines the best of the horizontal and vertical approaches.

Figure 5.7   Vertical components: seven difficult APIs.

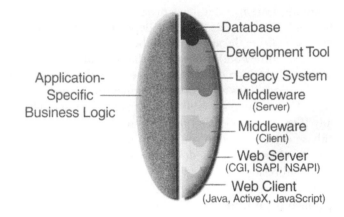

Figure 5.8   Business objects: a series of vertical components.

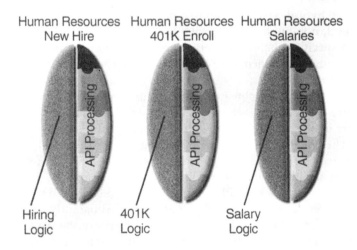

Using the Hybrid Approach

Most vertical components have similar API calls. To take advantage of this, you could define a horizontal component as an API across all vertical components, as shown in Figure 5.9. Processing is contained within the horizontal components and business requirements are contained within the vertical components. Unfortunately, other than some lines on paper (and in this case, a fancy picture), we still haven't built anything.

Figure 5.9   **Horizontal components: encapsulate APIs.**

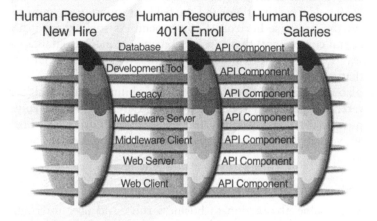

Next, you can extract the horizontal components into their own space. By providing a consistent set of wrapper functions for the horizontal components, all horizontal components are reduced to a single API. This removes, for example, the details of interprocess communication from the vertical components, letting you concentrate on business logic instead of communications. Essentially, the processing has been reduced to one API. Of course, someone has to build this API. I recommend your best developers do this, because the efficiency and usability of your application depends on the robustness of this API.

For horizontal and vertical components to work together, horizontal components must be intelligent — that is, they must ascertain information from specific vertical components and decipher what the vertical components require. To better understand this scenario, consider the example of a database API horizontal component. If a two-tier client/server developer wanted to access the database, he or she would either write SQL or have a development tool write it from query by example or a screen-based wizard. In a multitier environment, however, the actual database interface code is on the server, not the client where most client/server tools reside. You could still design the database horizontal component to accept a SQL statement as a parameter, but some of your tool's advantages will be lost. Thus, two-tier tools are not a good solution in a multitier world. Instead, you should let the horizontal components cooperate and exchange data with each other.

The first step in using horizontal components is to create a screen with fields. The screen must also contain information on the relationship between those fields and the tables whose columns they represent, as well as any foreign key information from the database schema (essentially, a query tree). You can obtain this meta data (data about data) by querying the database and saving the information in a meta data store or repository. When the user requests an event (for example, retrieve data for viewing) the user interface horizontal component first protects all fields, then passes the field list, the query tree, and the user event ("select" in this example) to the database horizontal component.

Next, the database horizontal component dynamically constructs the SQL statement required and returns the results to the user interface horizontal component. If another database event had been chosen (such as "insert"), the actions of the application code would be identical: the invocation of a database access. The user interface horizontal component would then get values from all fields, passing them and the field list, query tree, and user event ("insert" in this example) to the database horizontal component. As such, there is a common code (the database horizontal component) that is used in every instance of database access. Because this principle applies to the other horizontal components, we have essentially made all of the processing code (that isn't business rule-specific) reusable.

You must also consider horizontal components' ability to deal with unusual situations or factors that weren't apparent at the time they were created. To be truly robust, horizontal components must be extensible globally and for specific transactions. For example, in a specific transaction, you might want to modify the component-generated SQL but leave the overall model intact.

More than the Database Interface

You can apply the same abstraction to business rules and user interfaces by removing the business rules to a repository and the widgets to a class library (or better still, the same repository). In Figure 5.10, the vertical components inherit their processing from the horizontal components. They get their look and feel from the class library, and their business rules from the repository. The repository captures the application's meta data, including database schema, business rules, widget customization, and custom graphics — everything needed to build vertical components and provide information to the horizontal components. The repository entries are an example of multiple inheritance, an object-oriented concept. A repository object inherits its look and feel from the widget library of the platform on which the application runs, its meta data from the database schema, and its business rules from the developer.

Figure 5.10   Solution: multiple inheritance.

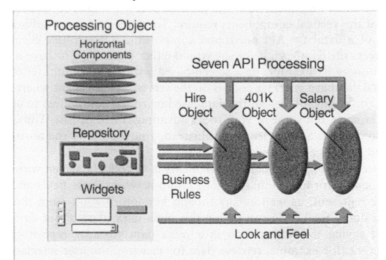

You can also perform an additional abstraction on the horizontal components. If you define a processing object as the set of horizontal components, as shown in Figure 5.10, the vertical components (or applications) then deal with only a single object for all processing. The horizontal components within this processing object communicate with each other in addition to communicating with the vertical components. This permits transparent error processing and recovery as a side benefit. If properly designed, the processing object can be reused across projects and even across business units. Finally, reuse as it was intended!

This approach has further benefits. Since the objects in the repository inherit their look and feel from the platform, the same application can run on different workstations and will take on the appropriate user interface for that platform. (Assuming, of course, you include the means to map each workstation platform into the repository). Thus, you can use the same application on the World Wide Web (using browser controls), any flavor of Windows, Macintosh, Motif, or even character-mode terminals. The vertical components are unchanged across platforms.

Similarly, the repository can encapsulate the database interface, so applications can be independent of the specific database engine chosen. Again, the vertical components are unchanged if the database engine changes or if multiple heterogeneous databases must be accessed within the same vertical component.

You can also extend the environment without actually modifying the vertical components. For example, suppose you discover the need for an added security layer such as data encryption. You can create a new horizontal component and add it to the processing object. Although, you may have to modify the adjacent horizontal components, the addition is often transparent to the vertical components that use the processing object. When the vertical components are affected, you can frequently make the changes by adding a property in the repository that is then distributed to the vertical components.

But what overhead is incurred in all of this sophisticated processing? The answer is not simple. In most applications, the real reasons for poor performance are underused or misused technology or poor design, rather than issues related to tight code. But poorly designed horizontal components can cause a performance hit. As mentioned earlier, it is critical that top-tier developers construct the horizontal component infrastructure.

The chief drawback to this approach is that the methodology must be in place before a project begins, through the design of the horizontal components and processing object. It also requires a team of experts to maintain the horizontal components who will most likely need to be paid more than the average programmer.

Transparent Middleware

I haven't addressed the question of communications between the layers. This "messaging middleware" is critical in multitier applications, but is difficult to learn and use. However, the service that processes the business rules is all you need if you can make the middleware and database horizontal components smart enough. I've already shown how you can do this for the database component. The messaging middleware layers also need some additional intelligence to make them transparent.

The trick is to make a common data store available to the horizontal middleware components that describe the vertical components. You can access this common data store from both the client and server hardware. It lists the vertical components and the inputs and outputs of each. Thus, all a "client" within a vertical component has to do is call the processing object requesting service. This middleware interface file can reside on a server (or multiple servers for redundancy) and is accessed through the middleware from both clients and servers. The horizontal middleware component on the client accesses the interface file, identifies the service and arguments it needs, and invokes that service, extracting the required data from the client. The server horizontal middleware component accesses the same interface file to determine the characteristics of the client invoking it. As such, it knows what data to expect and to return.

The amount of code savings you can achieve with effective use of vertical and horizontal components is substantial. For example, each client or service in an application typically requires 100 or more lines of API code that just manage the messaging. This includes code to manage message buffers, encode and decode the data to be passed, and process errors. You can replace all of this with a single statement call to the middleware horizontal component on the client and a single command in the service to receive the message and parse it automatically. Similar savings are achieved with the database horizontal component; all you need is a list of fields and the desired operation (View, Update, Rollback, and so on). The horizontal database component uses this information and the meta data stored in the interface file to build the SQL statements, then it makes calls directly to the database server by using the server's API.

Putting It All Together

The current state in multitier software development is not achieving expected gains. Therefore, new, yet simplified development methods must be invoked to achieve the successful development of timely and cost-effective enterprise-quality systems.

One way of getting notable improvements in new application development productivity is a hybrid approach using both horizontal and vertical components. Although not the strict definition of multiple inheritance as it pertains to objected-oriented development methodologies, this approach carries a similar concept. Imperative to the process is building intelligent horizontal components that query data structures in the application's vertical (business-oriented) components.

Although the principles I've discussed aren't easy to implement, the simplifications they provide for developing applications are well worth the investment in development effort.

Horizontal Components and Enterprise JavaBeans

Recently, Java has been expanded into the component model with JavaBeans and Enterprise JavaBeans. How do these technologies fit in with the horizontal component model?

JavaBeans are prepackaged application elements you can incorporate into a Java application. They are usually controls (pushbuttons, text boxes, and so on), but they can be more complex and incorporate application business logic. They are similar in concept to ActiveX controls; however, they are portable across all platforms that support Java. JavaBeans are a significant step in speeding the development of client Java applications and in making Java more competitive with other front-end tools.

The Enterprise JavaBeans specification is the model used to develop the server side of client/server systems. Enterprise JavaBeans server components execute within a component execution system — a set of run-time routines for the server components providing services such as threading, transaction management, state management, and resource sharing. The key element is that the specification defines how the applications interact with the component execution system, but not how the component execution system implements its services. You can adapt systems such as transaction processing monitors, CORBA or COM run-time systems, database systems, web servers, and so on to work in the Enterprise JavaBeans environment.

The component execution system of Enterprise JavaBeans is an implementation of horizontal components for the server side of a client/server system. It fulfills many of the objectives of horizontal component environments, but doesn't go as far as the horizontal component model in that it does not extend to the client. Using Enterprise JavaBeans, you can create services to be used by the client application. You can also define a standard interface for the client. The client must still be coded in Java (using native Java and JavaBeans) or in some other development environment; it must then invoke the services created with Enterprise JavaBeans. On the server, however, Enterprise JavaBeans provides the same environmental transparency to the application's vertical components as the horizontal component model.

In a fully implemented horizontal component environment, a horizontal component would manage the client-service interface, and horizontal components on the client side would manage the user interface, transactions, concurrency, security, and so forth. This arrangement helps you create true vertical components that are independent of the run-time environment from the user interface through the data access.

5.3.4 "Architecting for Change"

by Scott W. Ambler, May 1999

Be prepared for change long before it happens by implementing change cases in your design.

Change happens. Any successful software product will need to change and evolve over its lifetime. Extensibility, software's responsiveness to change, has been touted as one of the great benefits of object orientation, component-based development, and virtually any development environment you've ever worked with. It's a wonderful marketing spin, but in reality, software is extensible because it is designed that way — not because you implemented objects, not because you implemented components, and certainly not because you used the latest version of a certain product. Now, all you need is a technique to help you design extensible software. Enter change cases.

You can use change cases to describe new potential requirements or modifications to existing requirements for a system. Change cases are similar in concept to use cases: whereas use cases describe behavioral requirements for your system, change cases describe potential requirements that your system may need to support in the future. It is important to note that you can apply change cases to both behavioral and non-behavioral requirements, although in this column I will focus on an example of behavioral requirements. In his book *Designing Hard Software: The Essential Tasks* (Manning Publications Co., 1997), Douglas Bennett describes in detail how to use change cases to make your system more robust. I highly recommend this book.

You should develop change cases as part of your overall modeling efforts, during requirements gathering, analysis, and design. Change cases are often the result of brainstorming by your modelers and key users, who explore what technology can change, what legislation can change, what your competition is doing, what systems you will need to interact with, and who else might use the system and how.

You can document change cases by describing the potential change to your existing requirements, indicating the likeliness of that change occurring, and discussing the change's potential impact. In many ways, a system's change cases entail a well-documented risk assessment for changing requirements.

Consider the following example. Figure 5.11 depicts a use case model for a simple customer sales application. There are two use cases, Create Customer Order and Ship Order, that describe how to create and fulfill orders respectively. There are three actors, Sales Person, Customer, and Time. You can use Time to initiate an action. For example, if an order is scheduled to ship on a certain date and time, Time can invoke the use case to ship the order at the appropriate moment. This is a modeling trick I picked up from Gary Evans, an object development consultant based in Columbia, S.C.

Figure 5.11   Use case model for a sales application.

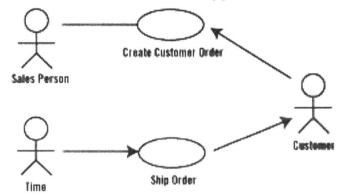

State transition diagrams don't address this issue of time. They show how an object will change over time, but they don't show time itself. For example, a state transition diagram for a bank account would show that accounts can be active, overdrawn, or closed. It would also show that accounts move back and forth between these states.

Note that in Figure 5.11 some of the relationship lines have arrowheads. In the UML, arrowheads indicate the entity, either an actor or a use case, that initiates interaction with the other entity. For example, Customer initiates interaction with the Create Customer Order use case, which in turn initiates interaction with a Sales Person. The lines in a use case model represent interaction, not data flow. The use of arrowheads is optional in the UML, and I prefer not to use them, as many people find them confusing.

Now consider some change cases for this system, as shown in Figure 5.12. I've applied the UML use case notation for the change cases, since the UML does not include the concept of change cases yet. I chose to create a separate model, although you could easily include change cases in your use case model. If you do, you should also mark each change case with a UML stereotype, something like <<Change Case>>. A stereotype, indicated with the <<some text>> notation, is a UML mechanism for extending the modeling language by defining a common type that is appropriate to your problem domain.

Figure 5.12   Change case model for a sales application.

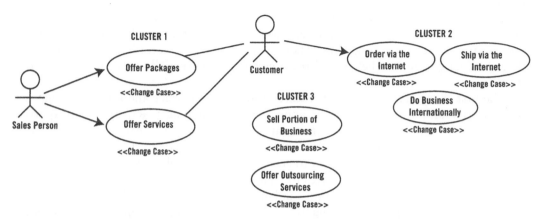

The advantage of including change cases in your use case model, or in any require-ments-oriented model, is that you can model dependencies between the change cases and por-tions of your model that are potentially impacted. The disadvantage is that this quickly clutters up your diagrams, which are probably complex enough at this point.

There are three clusters of change cases in Figure 5.12. Cluster one, made up of Offer Packages and Offer Services, describes simple changes to your business. Offer Packages describes the concept of offering collections of products, such as offering a large pizza, bread sticks, and a soft drink as a single package for your customers. The second change case indi-cates that your organization may offer services to your customers, such as delivery, as well as sell them products.

To document the Offer Packages change case, you must describe the potential change, which in this case you could do in a couple of sentences. You then document the likeliness of this happening. Assuming it's likely, you can indicate the time frame when it could happen, in this case about an hour. Finally, you must describe potential impact.

Your screen design will probably change, since your sales people will need to find out what products make up a package. You would also need to modify the administration system that your users use to define the products and services your organization sells. Further, your design would be impacted, so you would probably need to apply the Composition design pat-tern to implement products instead of having a single "Product" class. Knowing this, you might implement the concept of products as a domain component that in turn is built from one or more classes.

Now consider the Offer Services change case. The implication here is that your organi-zation may consider selling services in addition to physical products. You will need to imple-ment the services somehow, likely through a combination of software and the work of one or more people. There may be changes to the user interface for your application, depending on the information you would need to collect for each type of service. This change case also sup-ports developing a domain component, although it would be called something like Offerings because it would need to represent both the products and services that your organization offers. This implies that one change case may not be compelling enough for you to make a change to your design, although often a combination of change cases will. It would be wise to develop the Offerings component right away, but you should only put in the behavior you need for selling single products, making your design robust enough to easily handle these changes.

Cluster two — comprised of `Do Business Internationally`, `Order via the Internet`, and `Ship via the Internet` — concentrates on changes that would help expand your customer base via electronic commerce. These change cases are interesting because they not only describe changes to your behavioral requirements but potentially to your technical requirements.

The first change case is likely to require serious user interface and internal design changes to support multiple languages, your approach to calculating taxes, and your approach to shipping. It would probably motivate a data-driven approach to displaying text in the user interface, the use of a double-byte character set such as Unicode to store information, and the development of a domain component to encapsulate tax calculation. The two Internet-related use cases also describe serious changes to your system.

Cluster three describes two major, fundamental changes to your business, `Sell Portion of Business` and `Offer Outsourcing Services`. Because it is common for corporations to sell divisions as well as take over specialized portions of other businesses, designing your software to enable these sorts of business changes can be beneficial. These change cases, and several others, are likely to affect more than just your project and should be shared with other project teams in your organization.

Change cases offer several advantages. First, they let you consider and document possible changes to your software, justifying your efforts to develop a more robust solution to your existing problems. Second, they get your team out of the common short-term mentality, a mentality that generally leads to project failure (remember that the software industry has an 85% failure rate for large-scale, mission-critical projects). Third, change cases support incremental delivery of software by giving you a head start on the requirements for future releases and by increasing the chance that your existing release will support these additional requirements.

Finally, including change cases development in your software process ensures that your project team considers, and then documents, potential modifications to your system. If your team members don't consider potential extensions, the software they develop will probably not be easy to extend. In short, change cases lead you to develop software that is more maintainable, portable, modular, extensible, usable, and robust.

Change is an inevitable, fundamental aspect of our industry. Everybody recognizes this fact, yet for some reason a majority of us don't manage change or plan for it in our designs. We've always had excuses for getting caught off guard by change — users keep changing their minds, users don't know what they want, new legislation was introduced that affects business processes, or the marketing people are insane (O.K., that might be a valid excuse). Depending on the current software's design, it is typical that some changes result in "easy fixes." The majority, however, usually require significant rework to existing software, and some changes you cannot support economically. Change cases are a simple technique for minimizing the impact of change by preparing for it long before it happens.

5.3.5 "Enterprise Application Integration From The Ground Up"

by David S. Linthicum, April 1999

Choosing middleware to suit your specific integration needs is a step in the right direction. This article, the first of a three-part series, explains what the different types can do for you.

Need to tie more than one system or data source together? Then you need to know more about enterprise application integration (EAI). EAI is not only the latest acronym from the press and analyst community, it's a problem developers have been trying to solve ever since we began moving applications off centralized processors.

Chances are, if you're working in an enterprise, you have generations of systems using all sorts of enabling technology that have been built over the years. These may include mainframes, UNIX servers, Windows NT servers, and even proprietary platforms you've never heard of. What's more, the popularity of packaged applications — such as SAP, PeopleSoft, and Baan — has intensified the need to integrate those systems with other custom and packaged systems.

EAI is really a backlash to traditional distributed computing. Now that generations of developers have built these automation islands, more users and business managers are demanding seamless bridges between them. While the business case is clear and easy to define, this task is not.

In this article, and in the next two articles of this three-part series, I'm going to take on the EAI problem, working from the platform and network on up to the application and data, as shown in Figure 5.13. I'll look at low-, middle-, and high-end solutions that may be just the technology you're looking for. I'll also look at approaches to using technology, and how approaches vary according to the applications and the environment.

Let's begin our journey into the EAI world by examining traditional middleware technology, including message-oriented middleware, synchronous middleware, transaction middleware, database middleware, and distributed objects. I'll examine the technical features of each type, its application to EAI, and the pros and cons of each.

What's EAI?

So why should you care about EAI? It's something you've been spending a lot of time on over the last several years, without ever giving it a proper name. In fact, Forrester Research estimates that developers spend up to 35% of development time creating interfaces and points of integration for applications and data sources outside of their control. Moreover, most problems that pop up when developing software have to do with integration. Certainly it was the number one problem when traditional client/server systems were created. What was cheap to build is expensive to integrate.

EAI is unrestricted data sharing among any connected applications or data sources in the enterprise. You need to share this data without making sweeping changes to the applications or data structures. In other words, you must leave it where it lies.

Figure 5.13   To understand EAI, look from the network up to the application and data.

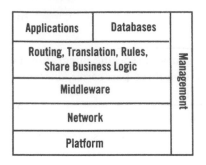

EAI's business value is obvious. The development of the islands of automation over the past 20 or so years has caused the EAI problem. For generations, developers have been building systems that serve one purpose and one set of users. Your enterprise probably has instances of these stovepipe systems. Some are inventory control systems, sales automation systems, general ledger systems, and human resource systems. Typically, these systems were custom-built using the technology of the day, much of which was proprietary and may have used non-standard data storage technology.

Despite the fact that the technology is old, the applications still have value and may be critical to the enterprise. Many of these business-critical systems are difficult to adapt to let them share information, and the cost of integrating them may be prohibitive. Many run on mainframes which, contrary to recent reports, are actually growing in use.

Lately, developers have added packaged applications such as SAP, Baan, and PeopleSoft, which are natural stovepipes unto themselves. Sharing information among these systems is a problem, since many were not designed to access anything outside their own closed proprietary technology. However, under the watchful eyes of screaming users and administrators, many ERP vendors are creating APIs or other integration points to let you access both methods and data from their packaged applications. But don't count on these APIs interoperating overnight.

Making the Connections

Truthfully, developers have been doing application integration for years, using all types of connection technology. In the past, this has been typically a low-level play with developers working at the network protocol layer or just above. More recently, developers have been using sophisticated middleware products such as IBM's MQSeries, BEA's Tuxedo, or paying homage to one of the two distributed object models: Distributed Component Object Model (DCOM) and Common Object Request Broker Architecture (CORBA). However, even these technologies have architecture and scalability limitations when connecting many systems together. There is still no solution to the EAI problem, despite what technology vendors may claim.

You are now facing the next generation of middleware with new categories such as message brokers, application servers (this one is actually old, but indulge me), and intelligent agents — many of which are simply not ready for prime time. This new middleware, however, must function in a business-critical environment. This means it must support continuous operation without fail.

Middleware hides the complexities of the underlying operating system and network to make integrating various systems with an enterprise easier. You deal with each system's API, and the middleware passes information through the different systems for the application. These APIs, however, are general-purpose information and data movement mechanisms, and typically don't know what applications and databases they are tying together. Developers have to program the links with applications and resources.

Middleware is an easy way to get to external resources using a common set of application services such as an API. External resources may include a database server, a queue, a 3270 terminal, ERP applications, or access to real-time information. In distributed computing, middleware is usually a means of connecting clients to servers without having to negotiate through many operating systems, networks, and resource server layers. However, peer-to-peer middleware is on the rise as a solution to the EAI problem.

In the early days of distributed systems development, you had to employ protocols directly to access either information or processing services on other computers. Therefore, you had to commit to a particular protocol, which limited compatibility with other platforms and networks.

Middleware's networking components function like a protocol converter. The distributed systems developer uses an API to dispatch a message. The network services translate the message into the appropriate protocol (for example, TCP/IP, HTTP, or NetBIOS), where it gets transmitted to another computer. The resolution of protocol differences happens in the background, away from the application. Using this model, you can conceivably swap network protocols, even platforms, every day, without making changes to the applications and without the users ever knowing if the API does not change.

Besides protocol translation services, middleware also translates data on-the-fly between mainframes, UNIX systems, and Windows machines. While this can happen in the background as more of a low-level data translation service, or simply platform translation, middleware has been moving to higher-level data translation services lately. Message brokers, one type of middleware, can alter the structure of a message so it makes sense to the application or database that's receiving it.

The arrival of advanced middleware layers let distributed application development take off. This newer middleware lets developers and application architects tie many different systems together to form virtual systems. In fact, according to analyst groups such as Gartner Group, middleware is moving to a "Zero Latency Enterprise" paradigm. Zero Latency Enterprise is where any application (or transaction) has access to another application or data store instantaneously, no matter what the protocol. This is a goal worth shooting for, whether or not it's achievable.

Although there are a variety of products and proprietary middleware solutions with differing features and functions, middleware generally falls into six separate types: remote procedure calls (RPCs), message-oriented middleware (MOM), transaction processing (TP) middleware, database-oriented middleware, object request brokers (ORBs), and the newest category, message brokers.

Remote Procedure Calls

RPCs are simply a method of communicating with a remote computer where you invoke a remote procedure on the server by making a simple function call for the client. RPCs hide the intricacies of the network and operating systems via a function call. The client process that calls the remote function suspends itself until the procedure is complete. This means that RPCs are synchronous and block the process from continuing until the function returns.

Although RPCs are easy to use and understand, they can be hard to incorporate into modern distributed computing. First, stopping an application from processing during a remote call could hurt an application's performance, making the application depend on the remote servers. Second, the RPC's architecture requires a high-speed network since clients and the server require a tremendous amount of network communications when using RPCs. Therefore, RPCs are typically only good for high-speed networks. Moreover, RPC is the only middleware that's actually declining in use, although both CORBA and DCOM leverage RPC-like communication mechanisms.

An example of an RPC-based middleware layer is the Distributed Computing Environment (DCE), developed by the Open Software Foundation and resold through vendors such as IBM/ Transarc. DCE makes many diverse computers work as a single virtual system. But DCE is a classic RPC mechanism with all of RPC's inherent limitations, including blocking (RPCs block processing until the procedure call returns) and the need for a high-speed network. Further, DCE uses a large amount of resources and suffers from chronic performance problems that are difficult to work around. While there is a place for synchronous middleware such as RPCs, you can count on RPCs becoming part of future operating systems and thus a commodity or, at least, common denominator service.

Message-Oriented Middleware

If you don't have the bandwidth to support RPCs, and if you can't depend on a server always being up and running, MOM may be a better fit for you. Like RPCs, MOM provides a standard API across hardware and operating system platforms and networks. MOM can also guarantee that messages reach their destination, even when destinations are not available.

MOM uses one of two models: process-to-process messages or message queuing. The process-to-process model requires that both the sending and receiving processes are active to exchange messages. The queuing model stores messages in a queue, so only one process can be active at a time. The use of the queuing model makes sense when communicating between computers that are not always up and running, over networks that aren't dependable, or when your system lacks bandwidth.

The basic difference between MOM and RPCs is that MOM is asynchronous. This means that MOM is non-blocking; for example, it does not block the application from processing when invoking the middleware API.

MOM message functions can return immediately even though the request has not been completed. Thus, the application can continue processing, assured that the application will know when the request is complete. This model is most useful for transaction-oriented applications that need to traverse many platforms. However, the MOM product must be able to maintain states to function properly.

Unlike DCE, a particular platform does not necessarily have to be up and running for an application to request services. The request sits in a queue, and when the server comes back online, the request is processed. This makes MOM a better fit for the Internet, since the

Internet does not have a lot of bandwidth and many Internet servers may not always be up and running. Moreover, this also makes MOM good for EAI, since many of the connected systems may or may not be up and running and MOM doesn't depend on a reliable network connection.

MOM's advantage is the flexibility of a message-based system. Since everything is a message, and messages are simple mechanisms (such as e-mail), you can adapt it to most applications. Moreover, MOM's asynchronous nature lets applications continue dispatching messages whether or not another application is ready to receive them.

MOM is becoming more advanced with features such as persistence support and the ability to maintain states. Persistence means that messages are stored to disk and a computer crash won't cause messages in transit to get lost. Maintaining states means that MOM can become the conduit for transaction-based systems. In other words, MOM will be able to maintain distributed transaction integrity.

Transaction Processing

TP monitors are industrial-strength middleware products that provide many features that make large-scale distributed transaction-oriented development possible. TP monitors are still the only way to go for high-volume distributed systems. A holdover from the mainframe world, TP monitors are a sound architectural solution for scaling distributed systems to an enterprise-level system (1,000 client/server users or more).

While TP monitors such as BEA's Tuxedo, Microsoft Transaction Server (MTS), and IBM's CICS are all but household words, there is a new breed of middleware known as application servers. Like TP monitors, application servers support the notion of a transactional system. WebLogic, Netscape Application Server (formerly Kiva), and NetDynamics are some of the new application servers. In addition, many traditional development tool vendors and object database vendors are moving into the new application server space.

Message brokers are middleware for middleware. They let you use a variety of enabling technology and integration points to tie many systems together.

TP monitors manage transactions for the client. They can route transactions across many diverse systems, as well as provide load balancing and thread control. You can even assign execution priorities to transactions to better manage performance, or launch transactions on a predetermined schedule. TP monitors and application servers add value to EAI, since you can use the TP monitor environment to define common shared application logic, letting one or many applications share both processing logic and data. To do this, however, you must tightly couple the systems, and make significant changes to the source and target systems that share information. TP monitors provide an independent location for business processing. The client acts only as an interface to the TP monitor, invoking functions. The TP monitor, in turn, accesses the database, or other resource server, for the client.

One of the most significant features of a TP monitor is its ability to multiplex and manage transactions. This removes the one-client-to-one-connection restriction of traditional two-tier client/server development. The TP monitor can only use a few connections to a database server to manage hundreds of clients. This is known as database funneling. Database funneling means that you can use a TP monitor to create client/server systems that handle more clients than the database server could handle comfortably. If the number of clients increases, the TP monitor simply kicks off new share connections to the database server.

TP monitors also protect distributed applications from potential disasters, since they can reroute transactions around server and network failures. In addition, TP monitors can recover if a transaction fails. They never leave the system unstable. They can use different connecting middleware types, including RPCs and MOM, and can even communicate with database servers using native database middleware or ODBC.

Leveraging the power of the transaction is the main advantage of TP monitors. Since transactions are all or nothing, TP monitors can ensure that critical business events take place. There is no in-between, and application and database integrity is protected. The database-multiplexing trick is an advantage as well, and lets distributed architectures scale since the database server is no longer the limiting factor for scalability. The disadvantage is that TP monitors force you into the transaction paradigm, no matter what type of processing you carry out. Everything has to be bundled in a transaction.

The future of TP monitors is application servers. Traditional TP monitors are finding it difficult to keep up with the web-enabled wonder servers available from WebLogic, NetDynamics, Novera, and many others. These application servers are easier to use and program and possess many of the same features as TP monitors.

Database-Oriented Middleware

Database-oriented middleware, by contrast, refers to middleware built specifically for database access. Database-oriented middleware is the software that connects an application to a database. It lets you access the resources of a database server on another computer, using a well-defined API. Although database middleware is easier to understand architecturally, there are many products in this market, and they all work differently.

SQL gateways are single API sets that provide access to most databases on different types of platforms. They are like virtual system middleware products, but are specifically for database processing. For example, you can use an ODBC interface and an SQL gateway to access data in a DB2 on a mainframe, Oracle running on a minicomputer, and Sybase running on a UNIX server. You simply make an API call, and the SQL gateway does the work for you.

SQL gateways translate the SQL calls into a standard format known as the Format and Protocol (FAP). FAP is a common connection between the client and the database server, and is the common link between different databases and platforms. The gateway can translate the API call directly into FAP, moving the request to the target database. It then translates the request so that the target database and platform can react.

Database-oriented middleware is essential with EAI since most of the enterprise data exists in databases that are only accessible using this type of middleware. Simply accessing the data, however, does not solve the EAI problem. The data must be distributed, identified, classified, and altered to reach the proper application that requires the data. That's where middleware like MOM comes in handy, and where message brokers are proving to be a better type of middleware.

A main advantage of database-oriented middleware is that developers have simply been using it for so long that we're good at it. Typically, database-oriented middleware is now built into the front-end programming tools, although many developers may not know it's there. That's certainly the case with some of the more advanced Java development tools, such as

Inprise's JBuilder, IBM's Visual Age for Java, and Microsoft's J++. They expose the database schema to you, but not always the API.

The disadvantage of database-oriented middleware, such as call level interfaces (for example, ODBC and JDBC), is that it forces you away from the database, so it's difficult, if not impossible, to leverage native features of the database including stored procedures, database objects, and triggers. You must use a common API that does not leverage the special features of a particular database.

Message Brokers

When you look at what message brokers bring to business, their value is clear. A message broker can act as a broker between one or more target entities (such as network, middleware, applications, or systems), letting them share information with less pain than using traditional middleware. In other words, a message broker is middleware for middleware. It lets you use a variety of enabling technology and integration points to tie many systems together.

Message brokers facilitate the integration of business processes, which until now have been more isolated than open. These message brokers use a hub and spoke-type architecture (as shown in Figure 5.14) and can "broker" information (messages) among any number of entities (applications), no matter how you represent or access the information.

Figure 5.14   The message broker architecture.

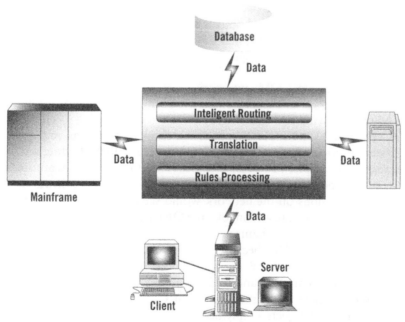

The message broker's purpose is to integrate multiple business activities (applications), whether they are new, old, legacy, centralized, or distributed. This technology ties together many different platforms and application development solutions, and it can also connect to each application using a variety of middleware and API mechanisms and route information among all of them.

In addition to simply routing information, a message broker may provide enhancements such as letting message brokers host business functions that build on the business functions of the entities to which they connect. For example, you may build a data extractor for payroll that can also identify potentially fraudulent information coming out of the payroll system. Although it can enhance business functionality, a message broker's primary role is to provide a simple, central point of integration.

Message brokers often provide data translation engines, which let the message brokers change the way information is represented for each application. Considering the Year 2000 problem, for instance, the date 1998 may be represented in a legacy system as "98" and in a packaged application as "1998." You can apply the same principle to currency data representation within many applications, and even to how you might represent customer numbers across enterprise applications. I'll examine the concept of the message broker in more detail in the next two articles.

Message brokers have the advantage when it comes to integrating many systems together, and thus, are natural for EAI. Many-to-many integration is something that neither RPC nor MOM can do, since they are traditional point-to-point, or single-to-single applications. The downside, however, is the immaturity of the message broker industry. Most of today's message brokers and vendors don't have the long-term experience with the enterprise, and many of them fall short in features and functions to support enterprise paradigms such as EAI.

Distributed Objects

As I mentioned earlier, there are two models for distributed objects: DCOM and CORBA. While it's difficult to call distributed objects middleware, they can provide many of the same features as MOM, message brokers, and TP monitors as they become more functional. This includes sharing data, application logic, and providing a central clearinghouse for enterprise information.

DCOM lets you create COM automation servers and make them available for other COM-enabled client/server applications on a network. DCOM is not a commercial ORB, but part of the Microsoft Windows operating systems.

With DCOM, the COM-enabled application simply checks the registry of the Windows operating system to locate and use remote COM-enabled ORBs, finding and invoking the service it requires. For example, you could create a COM ORB (an OLE automation server) that generates a sales report that any number of COM-enabled applications can use. Other COM-enabled applications on the network would see and use the COM Sales Report ORB. These applications could locate and invoke the ORB's methods through the Windows operating system's built-in DCOM mechanism.

DCOM differs from CORBA because it's built into the operating system's infrastructure, and does not come from an ORB vendor. DCOM is backward-compatible with existing COM-enabled development tools and tools that were not created specifically for distributed object development. These tools, such as Visual C++, Visual Basic, and Visual J++, will be used to create distributed objects simply because DCOM can distribute the COM automation servers they create.

CORBA-compliant distributed objects leverage Internet Inter-ORB Protocol (IIOP) as a wire between the objects. With the CORBA 2.0 specification, the Object Management Group (OMG) finally achieved interoperability between ORBs by defining the IIOP communications mechanism as mandatory for all CORBA-compliant ORBs.

Consider IIOP as TCP/IP with a few CORBA-specific messages for ORB-related communications. All CORBA-compliant ORBs must implement IIOP as a native feature of the ORB, or provide a "half-bridge" to IIOP. Half-bridges mean that any proprietary ORB can connect to other ORBs by linking to an IIOP backbone through a translation layer.

You can exploit IIOP's distributed features to create distributed applications. These applications can access a collection of ORBs connected via a network such as the Internet or an intranet. ORB vendors should be able to communicate with each other using the IIOP backbone, while proprietary ORBs can connect via half-bridges.

IIOP is a natural for EAI because of its distributed capabilities and native support for TCP/IP. The OMG standards body believes that EAI will raise interest in using distributed objects. Until now, few CORBA-compliant commercial ORB vendors have penetrated the enterprise software market. The OMG figures CORBA/IIOP and EAI will recharge the dying distributed object marketplace plagued by slow delivery cycles and a lack of tools.

Distributed objects let you share both methods and data using well-defined standard interfaces, and they have proven themselves as a mechanism to create distributed systems in recent years. However, the COM and CORBA debate has added perceived risk to the distributed object paradigm, and there are many technical limitations to overcome such as scalability and stability. Distributed objects will continue to serve a niche in the world of distributed computing, and interest in distributed objects will remain steady.

In this article, you learned about the basics of EAI as well as the middleware technology that enables it. In the next article, I'll discuss the new generation of middleware, such as message brokers, and how you can leverage this technology for EAI. Message brokers are clearly the preferred EAI engines for the enterprise, and the technology developers are likely to use most to solve EAI-type problems. I'll also cover some of the higher-level value propositions of EAI including the use of EAI as a mechanism to enhance the automation of business processes. You need to understand that ultimately EAI is more about enhancing the business processes than connecting systems together. This is the real value of enterprise application integration when all is said and done.

5.3.6 "Mastering Message Brokers"

by David S. Linthicum, June 1999

Message brokers are being heralded as the next generation of middleware. This article, the second in a three-part series, explains how they measure up to the challenges of enterprise application integration.

I covered enterprise application integration concepts, approaches, and an enabling technology overview in the previous April 1999 article (section 5.3.5 "Enterprise Application Integration From The Ground Up". In this installment, I will focus on the middleware technology that's likely to drive EAI for developers: message brokers.

In my previous article, I explained that a message broker acts as a broker between one or more target or source entities (such as network, middleware, applications, and systems), letting them share information more easily than traditional techniques.

Message brokers facilitate business process integration, which used to be more isolated than open. Most message brokers use a hub and spoke-type architecture to "broker" information (messages) between any number of entities (applications, objects, or databases) — no matter how you represent or access the information.

To this end, a message broker's purpose is to integrate multiple business activities (applications, objects, or databases), whether they're new, old, legacy, centralized, or distributed. This technology ties many different platforms and application development solutions together, and it can also use any number of middleware and API mechanisms to connect and route information to each application, as well as routing information between applications.

In addition to routing information, a message broker can enhance information routing by hosting business functions that build on the existing business functions of the entities they connect.

Message brokers often provide message translation engines, which let the message broker change the way each application represents information. Other message broker features include processing and applying rules to information translation and routing; intelligent routing; and identifying and routing messages to the appropriate location. Also, you can apply rules to certain message types and perform any number of operations, such as message translation.

You can separate message broker services into several distinct categories: message transformation services, rules processing, intelligent routing, message warehousing, flow control, repository services, directory services, APIs, and adapters. I will cover each of these services.

Integration, Not Perspiration

Message brokers are not new. Developers have been creating them in one form or another for years. The message broker vendors, including Active Software Inc. (www.activesoftware.com), New Era of Networks (NEON, www.neonsoft.com), Century Analysis Inc. (now a part of NEON), and TSI Software (www.tsisoft.com), are really just turning the concept into products.

In fact, many organizations, such as the U.S. Department of Transportation and Federal Express, have spent millions of dollars creating their own custom message brokers. They saw the need but not the product, so they created their own message brokers. You may have a first-generation message broker within your organization and not even know it.

Traditional middleware, such as message-oriented middleware (MOM), only solves part of the EAI problem. Message brokers do not replace traditional middleware like MOM, RPC, and distributed transaction processing monitors. Message brokers build on existing middleware technology, most often on messaging middleware. Therefore, you might call message brokers "middleware for middleware."

Many business integration solution components are neither middleware nor applications. These are typically routing, reformatting, and flow components. These components, which you can place in an application or in the middleware, are a better architectural fit within message brokers because they provide a central point of integration.

A message broker is based on asynchronous, store-and-forward messaging. It manages interactions between applications and other information resources utilizing abstraction techniques. Applications simply put (publish) messages to the message broker, and other applications get (subscribe to) the messages. Applications don't need to be session-connected, which eliminates the largest scalability problem of most integration technologies.

However, message brokers also extend the basic messaging paradigm by mediating the interaction between the applications so that the information publishers and subscribers can be anonymous. They translate and convert data, reformat and reconstitute messages, and dynamically route information to any number of targets based on centrally defined business rules applied to message content.

To serve as a solid foundation for an EAI strategy, message brokers must be truly "any-to-any" and "many-to-many." Any-to-any means connecting diverse applications and other information resources must be easy, the approach must be consistent, and all connected resources must have a common look and feel. Many-to-many means that once you've connected a resource and its publishing information, any other application with a "need-to-know" can easily reuse it. The publish/subscribe messaging model is one technique for addressing this, and its popularity is growing.

Message Transformation Layer

Many message brokers have a message translation layer that understands the formats of every message being passed among applications and changes their formats on the fly. The message broker restructures the data from one message into a new message that makes sense to the receiving application or applications. The message translation layer provides a common dictionary with information on how each application communicates and which information is relevant to which applications.

Typically, message translation layers contain parsing and pattern-matching methods that describe the structure of any message format. Message formats are constructed from parts that represent each field encapsulated within a message. If you break a message down to its components, you can recombine the fields to create new messages. There are two components of message transformation: schema conversion and data transformation.

Schema conversion is processing the changing message structure, thus re-mapping the schema to an acceptable state for the target system. While this is not difficult, EAI architects should understand that this process must occur dynamically within the message broker.

For instance, if a message comes in from a DB2 system on a mainframe with accounts receivable information from the accounts receivable system, it may look something like this:

```
Cust_No              Alphanumeric      10
Amt_Due              Numeric           10
Date_of_Last_Bill    Date
```

With the following information:

```
AB99999999
560.50
09/17/98
```

However, the client/server system, created to do the annual report, receives the data and must store it in the following schema:

```
Customer_Number    Numeric 20
Money_Due          Numeric 8
Last_Billed        Alphanumeric 10
```

As you can see, the client/server system and a DB2 system schemas are different, although they may hold the same data. If you attempted to move the information from the DB2 source system to the client/server target system, you would probably encounter a system error, since the formats are not compatible. You must transform the schema (and the data) so it works with the application or data store receiving the information.

Data transformation refers to the message broker's ability to change data on the fly, as it's being transferred between the source system and target systems. This means simple data conversion techniques, such as applying algorithms (as is the case with data aggregation), or using look-up tables.

Intelligent Routing

Intelligent routing (sometimes known as flow control) builds on the capabilities of both the rules and the message translation layer. The message broker can identify messages dynamically coming from a target application and route that information to the proper source application, translating the messages if required.

If a message comes to a message broker, it is analyzed and identified as coming from a particular system or subsystem — for instance SAP's Accounts Receivable module. Once the message broker identifies and understands its message's origin and schema, it processes the message. This means that it applies applicable rules and services, including message transformation, to processing the message.

Once it processes the message, the message broker, based on how it's programmed, will route the message to the correct target system. This may take less than a second, and up to 1,000 of these operations may occur simultaneously.

Rules Processing

The rules processing engine within most message brokers lets EAI architects and developers create rules to control the message processing and distribution. In simple terms, it's an application development environment supported by the message broker to support the special application integration requirements.

A rules engine will let you implement intelligent message routing and transformation. For example, in some instances a message may be required by one other receiving application. In other instances, you may have to route a message to two or more applications. To solve each problem, you should be able to route the message to any number of applications that are extracting and translating data from other applications.

Don't confuse the message broker rules processing capabilities with those of traditional application servers. Application servers provide full-blown application development environments and tools, whereas most rules processing engines only give you enough features to move messages between any number of source systems to any number of target systems.

Rules processing typically uses scripting languages rather than more complex programming languages such as Java or C++. What's more, they typically use interpreters vs. compilers. However, the way that message brokers process rules differs greatly from vendor to vendor and from message broker to message broker.

Message Warehousing

Another message broker feature is message warehousing. A message warehouse is a database that can store messages that flow through the message broker. In general, message brokers provide this message persistence facility to meet several requirements, including message mining, message integrity, message archiving, and auditing.

Message mining refers to using the message warehouse as a quasi-data warehouse, which lets you extract business data to support decisions. Message warehousing can provide services such as message integrity, since the warehouse provides a natural persistent state of message traffic.

If a server goes down, the message warehouse is a persistent buffer or queue to store messages that could be lost. You can use the message warehouse to resend or compare messages with other message warehouses on the network to ensure message transfer integrity. This is much the same principle as the persistent message queuing that traditional message-oriented middleware supports.

Repository Services

Today, many message brokers embrace the repository concept. Repositories are databases of information about source and target applications that may include data elements, inputs, processes, outputs, and relationships between applications. While most developers consider repositories part of application development, they have clear value to EAI.

The repository should provide all of the information EAI architects and programmers need to locate within the enterprise and link it to any other information. The repository is the master directory for the entire EAI problem domain.

The Microsoft Repository, now in its second year, and Platinum Technology's look-alike version will work fine with most message brokers. However, most message broker companies have initially chosen to leverage proprietary and more primitive repositories. This is primarily because the Microsoft repository supports application designers and developers, whereas message broker repositories track information moving throughout an enterprise.

Graphical User Interface

One great thing about a message broker is that it's middleware with a face, or at least a graphical user interface. Most message brokers come with a graphical user interface, or an interface that lets you create rules, link applications, define transformation logic, and so on.

Directory Services

Since message brokers deal with distributed systems, they must use a directory service to locate those systems. A directory service is a mechanism that provides the location, identification, use, and authorization of network resources, such as a source or target application.

The idea is to provide a single point of entry for applications and middleware (like message brokers), and to support using a shared set of directory and naming service standards. In other words, directory services help you find your way among the thousands of applications and middleware resources.

Adapters

The notion of message brokers adapters has been around for a while. However, each vendor has its own definition of an adapter, and to date, nobody can agree. But it's important to note that the capabilities of many adapters are grossly exaggerated.

Adapters are, generally speaking, a layer between the message broker interface and the source or target application. An adapter can adapt to the differences between the source or target and the message broker's native interfaces. It hides the complexities from users, or even the EAI developer using the message broker. For example, a message broker vendor may have adapters for several different source and target applications such as SAP, Baan, or PeopleSoft; adapters for databases, such as Oracle, Sybase, or DB2; or adapters for specific middleware brands.

In many cases, adapters are a set of libraries that can map the differences between two distinct interfaces, the message broker interface and the source or target application interface. For instance, when making a call to SAP, a message broker may leverage the adapter's features and functions to reach the services of SAP, without having to invoke SAP's native interface. The adapter lets you reach SAP's process and data levels, without creating an intermediary application to invoke the SAP application interface on one end and the message broker interface on the other.

In a sense, you're simply trading a complex interface for an easier one, hiding the complexities of the systems you're integrating from users who need to do the integration. Therefore, the user may only deal with the message broker's GUI; since the message broker uses an adapter, the user never deals with the source or target application's native interfaces.

Two types of message broker adapters are emerging: thin adapters and thick adapters. Thin adapters are offered by the most popular message brokers. Thin adapters, in most cases, are simply API wrappers or binders that map the source or target system's interface to a common interface supported by the message broker. In other words, it simply performs an API binding trick, binding one API to another.

Thick adapters, in contrast, provide a lot of software and functionality between the message broker infrastructure and the source or target applications. Thick adapters typically provide a layer of abstraction between the message broker and a source or target application, completely hiding all APIs from the user and making it painless to manage the movement or invocation of processes. This abstraction layer lets you manage the source or target applications with almost no programming, since the abstraction layer and the adapter manage the difference between all the applications you are looking to integrate.

Using an API

In addition to adapters, message brokers use APIs to access the message broker's services from an application. There is really nothing special here, and the message broker API looks very much like the API of a traditional message-oriented middleware product, except that the message broker lets you integrate many applications. Traditional message-oriented middleware, as you may remember, works best for linking one application to another.

Topologies

Message brokers typically use what is called the hub and spoke type of topology. This means the message broker, or hub, sits between the source and target applications, as Figure 5.15

shows. While this is more traditional, you can leverage message brokers using other topologies such as bus or multi-hub.

Figure 5.15   Hub and spoke configuration.

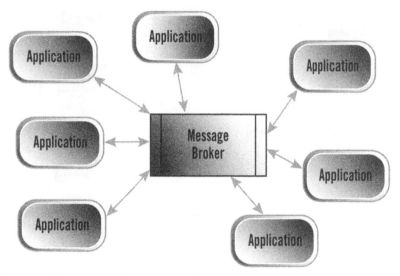

In the bus configuration, the message broker simply sits on the network bus, providing message broker services to other systems on the bus, as shown in Figure 5.16. This scenario works best when message brokers play less of a role in your EAI problem domain.

Figure 5.16   Bus configuration.

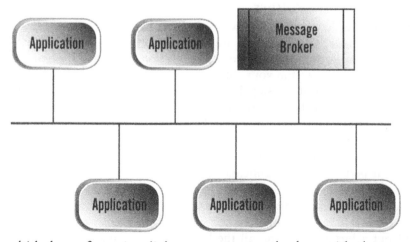

The multi-hub configuration links many message brokers with the source and target applications linked to any of the brokers in the configuration. Figure 5.17 shows the multi-hub configuration. The multi-hub configuration's advantage is the ability to scale. For example, you can integrate any number of source and target applications, because if you need to integrate more applications than one message broker can handle, simply add more message brokers to the network.

Figure 5.17   Multi-hub configuration.

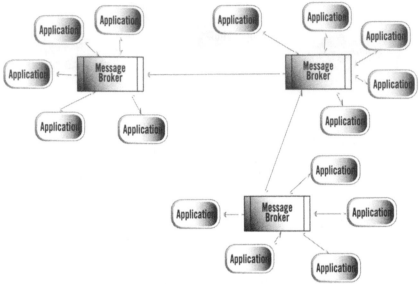

As message brokers begin to mature, they're moving from the simple hub and spoke type configuration to multi-hub with different degrees of sophistication. While some message brokers can support multi-hub configurations, they can't intelligently share the message processing load between the message brokers.

Future of EAI and Brokers

Is this the next generation of middleware? Indeed. Message brokers fill a need that many larger companies have today, and that traditional middleware won't solve. The downside is that you'll have to go through the normal confusion that arises when a new technology attempts to find its place within your enterprise.

If you've looked at most message brokering products, you know they all take very different approaches to solving the same set of problems. It may be a good idea for message broker vendors to get on the same sheet of music. Promoting the concept of a message broker — at least at first — could be more beneficial than promoting a particular product.

Message brokers, or at least their core engines, will become commodity products because many of them perform only rudimentary functions. To differentiate their products, many message broker vendors are adding layer after layer of value-added software to the core message broker engine. Thin adapters are giving way to thick, and many message broker vendors are learning that it is better to hide the complexities of enterprise systems behind abstraction layers and put a business face on all the little details. Ultimately, message brokers may be more about business automation than middleware.

In the third installment of this series, I'll wrap up my EAI discussion with a look at how all of this technology fits together to solve the EAI problem. What's more, I'll examine the general concepts behind EAI, and how developers fit into the mix.

5.3.7 "Building Enterprise Architecture"
by David S. Linthicum, September 1999

Before you begin consolidating your company's business logic, you must decide whether to integrate at the data, application programming interface, method, or user interface level.

So, what have I covered so far in this enterprise application integration (EAI) series? In the first installment (section 5.3.5 "Enterprise Application Integration From The Ground Up"), I explained the different types of middleware and where they work best. The second installment (section 5.3.6 "Mastering Message Brokers") examined how message brokers measure up to the challenge of EAI. While these technology components of EAI are significant, developers and enterprise architects must keep a much larger picture in mind. In this final installment of my EAI series, I'll look at some of the higher architectural issues of EAI, including approaches as well as enabling technology.

It's important to remember that EAI entails more than connecting applications. If that's all there were to it, I would discuss application programming interfaces (APIs), message queues, and other connectivity solutions that exist only at the programming level. EAI is actually a new approach to integrating applications, which may revolutionize the way you plan new application development or think about re-engineering existing applications. Reading this article is only your first step to mastering EAI.

Undoubtedly, there are a number of instances of non-integrated systems in your enterprise today, possibly for inventory control, sales automation, general ledger, and human resources, to name just a few. These systems were typically custom-built with a specific business problem in mind for a narrow set of users, utilizing contemporary development technology — such as the COBOL and DB2 movement of the mid-1980s, the UNIX and C revolution of the late 1980s, and the client/server trend of the early 1990s. To make matters worse, these systems were largely built without any notion of interoperability. Integration simply was not a requirement for the developers at the time, and EAI entails fixing these past mistakes.

While the technology has aged, the business value of the applications remains constant. Indeed, systems built using older technology have remained critical to the enterprises they serve. Unfortunately, it's difficult to adapt many of these business-critical systems to let them share information with other, more advanced, systems without a significant investment in time and money — not to mention risk.

Packaged applications such as SAP, Baan, and PeopleSoft — which are "closed applications" themselves — have only compounded the problem. Like custom enterprise systems, packaged applications were not designed to easily share information and processes.

EAI is a response to traditional enterprise application development and packaged application usage. The essence of EAI is unrestricted data and business process sharing among any of an enterprise's connected applications or data sources. While the goal of EAI is clear, its procedures and technology are more a science than a process.

The notion of simply sharing information between applications is easy enough to understand. However, many enterprises need to share both data and methods without making significant changes to the source or target systems. In other words, EAI must join applications at an existing point of integration, saving the expense and risk of having to change, test, and re-deploy applications. This is known as noninvasive EAI, and is the focus of most early efforts.

Of course, noninvasive EAI only applies to a select set of problem domains, and does not provide the infrastructure for reusing business logic, such as sharing distributed objects or transactions. However, to properly bind applications at the business process level, developers must make sweeping changes to the participating applications, letting them share common methods and thus data. This is, of course, intrusive EAI, and while the benefits are high, so are the risks and costs.

Implementing EAI

When contemplating EAI in your organization, you must first understand the sum and content of the business processes and data in your organization. Your staff also needs to understand how these business processes are automated (and sometimes not automated) and the importance of all business processes. Depending on your enterprise, this may demand a significant amount of time and energy. In addition, many organizations seek new methodologies to assist them in this process while looking closely at best practices.

To successfully implement EAI, organizations must understand both business processes and data. They must select which processes and data require integration. This process can take on several dimensions or levels of abstraction, including data level, application program interface level, method level, and user interface level.

Data-Level EAI

Data-level EAI means moving information between two or more databases in order to integrate the applications. You use data level EAI to extract information out of one or many databases, perhaps transforming its schema and content so it makes sense to the database receiving the data, and then placing the data in the target databases. While this may sound simple enough, more complex data-level EAI problem domains may entail moving data between hundreds of databases of varying brands and models. The idea here is to avoid changing the application logic while moving data between applications. The relevant technology is traditional database middleware, data movement software (such as products used in the world of data warehousing), application servers, and message brokers.

Data-level EAI provides simplicity and speed-to-market, and is typically cheaper to implement.

Accessing data in the context of EAI demands doing an "end run" around application logic and user interfaces, as shown in Layer 1 of Figure 5.18. Applications and interfaces were not designed to accomplish EAI, but rather to work independently. As a result, successful implementation of data-level EAI requires sneaking behind the application's logic and extracting or loading the data directly into the database through its native interface. Fortunately, most applications built in the past two decades or so decouple the database from the application and interface, making this a relatively simple task. However, this doesn't mean that data-level EAI is always a good idea. You must consider how tightly coupled the data is to the application logic. Moving data between databases without understanding the entire application is a dangerous maneuver.

Figure 5.18 EAI levels.

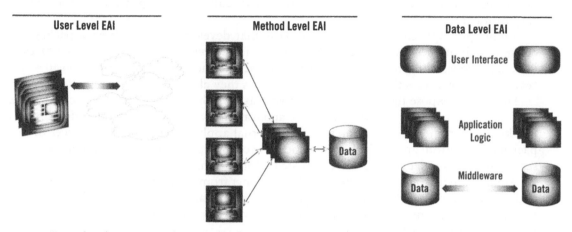

Data-level EAI provides simplicity and speed-to-market, and is typically cheaper to implement than other forms of EAI such as application interface, method, and user interface. What's more, you don't need to employ new and expensive EAI tools like message brokers when you've limited the scope of your EAI solution to moving information between databases. Simple data replication and transformation tools may provide all the power you need.

Cost and simplicity are clear advantages of data-level EAI because you rarely have to alter business logic. As a result, you don't need to endure countless testing cycles or the risk and expense of implementing newer versions of an application within any enterprise. Indeed, most users and applications won't know that data is being shared behind the scenes.

Given the task's conceptual simplicity and real life complexities, how does an enterprise implement data-level EAI? Ultimately, it comes down to understanding where the data exists, gathering information about the data, and applying business principles to determine which data flows where, and why. Many are compelled to create an enterprise-wide data model complete with meta data (data about data) to help EAI architects and developers better determine source data, as well as the potential target systems.

The numerous database-oriented middleware products that let architects and developers access and move information between databases simplify data-level EAI. These tools and technologies let you integrate various database brands, such as Oracle and Sybase. They also let you integrate different database models, such as object-oriented and relational models. However, if you're looking to move information from the data level to the application interface method or user interface levels, consider using more sophisticated EAI technology such as message brokers or application servers.

API-Level EAI

API-level EAI, like data-level EAI, avoids the cost and risk of having to change the source and target applications by using existing application program interfaces that might exist within custom or packaged applications. Using these interfaces, developers can cohesively bind applications together, letting them share business information and business logic, albeit in a loosely coupled but noninvasive manner. The only limitations that developers face are the application interfaces' specific features and functions, which vary from pretty good to nearly impossible to use. SAP R/3's Application Linking and Embedding (ALE) interface is an example of an API. However, it would be a stretch to call SAP an open application with an easy-to-use interface.

Simply put, application interfaces are those that developers expose from a packaged or custom application to access the application's various levels or services. Some interfaces are limited in scope, while many are feature-rich. These interfaces can let you access business processes or let you access the data directly. Sometimes they allow access to both.

Developers expose these interfaces for two reasons. The first is to let you access business processes and data encapsulated within the applications, without forcing other developers to invoke the user interface. Such interfaces let external applications access the services of these packaged or custom applications without actually changing the packages or applications. The second reason is to let you share encapsulated information.

Application interfaces as a type of EAI are distinct from method-level EAI (discussed next) and user interface-level EAI (discussed later). While you can distribute the methods that exist within enterprises among various applications, they are typically shared via common business logic-processing mechanisms, such as application servers or distributed objects. User interface-level EAI is similar to application interface-level EAI in that both make business processes and data available through an interface exposed by the package or custom application; user-level EAI simply does this through the existing user interface (for example, 3270, 5250, or OLE automation).

This approach distinguishes application interface-level EAI from other types of EAI. The potential complexities of the application interfaces, as well as the dynamic nature of the interfaces themselves, add to the difference. Because interfaces vary widely in the number and quality of the features and functions they provide, it is nearly impossible to know what to expect when invoking an application interface. Packaged applications are all over the map in terms of which interfaces they expose.

Packaged applications (which are most often present in a typical EAI problem domain) are only now opening up their interfaces to allow for outside access and, consequently, integration. While each application determines exactly what these interfaces should be and what services they will provide, there is a consensus in providing access at the business model, data, and object levels.

When accessing the business model, or the innate business processes, you typically invoke a set of services through user interfaces. For example, you can access credit information for a particular individual through the user interface by driving the screens, menus, or windows. You can also access this information by invoking an API provided by the packaged application vendor.

In the world of custom applications, anything is possible. With access to the source code, you can define a particular interface or simply open the application with a standard interface. For example, rather than accessing the user interface (scraping screens) to reach an existing COBOL application residing on a mainframe, you can build an API for that application and thus expose or extend its services. In most cases, this requires mapping the business processes, once accessible only through screens and menus, directly to the interface.

This approach's downside is cost. Half the costs go toward application changes and the subsequent testing and redeployment. In many cases, it is much cheaper (and just as effective) to simply access the application information by automating access to user interfaces from a program. This is, of course, user interface-level EAI.

If the world were perfect, all the features and functions provided by packaged applications such as SAP, PeopleSoft, Oracle Financials, and Baan would also be accessible through their well-defined APIs. However, the world is not perfect, and the reality is a bit more sobering. While almost all packaged applications provide some interfaces, they are, as mentioned already, uneven in their scope and quality. While some provide open interfaces based on open interface standards such as Java APIs (for example JavaBeans) or object request brokers (ORBs), many provide more proprietary APIs that are useful only in a limited set of programming languages (for example, COBOL and C).

Most disturbing, many packaged applications don't offer an interface. With these applications, an application or middleware layer can't access the services of that application. As a result, the business processes and data contained within the application remain off-limits.

Method-Level EAI

Method level EAI integrates applications by binding them together at the method level, as shown in Figure 5.18, Layer 2. This is, as described earlier, the most invasive form of EAI, requiring that major changes take place within all participating applications. We can do this through any number of traditional application method-sharing technologies such as distributed objects or application servers. This generally means creating a hybrid or composite application, which can provide the infrastructure for accessing shared business processes.

Attempts to share common processes have a long history, starting more than 10 years ago with the distributed object movement and multi-tiered client servers — a set of shared services on a common server that originally provided the infrastructure for reuse, and now facilitate integration. Reuse is important in this context. By defining a common set of methods, you can reuse those methods among enterprise applications. This significantly reduces your need for redundant methods and applications.

Absolute reuse, or the tight integration you get from reuse, has yet to be achieved on the enterprise level. The reasons for this failure range from internal politics to the inability to select a consistent technology set. In most cases, the limit on reuse is the lack of enterprise architecture and central control. So what does this failure of reuse have to do with EAI? The answer is: everything.

Using the EAI tools and techniques to integrate an enterprise not only helps you share common methods, but also provides the infrastructure to make such sharing a reality — integrating applications so that information can be shared and providing the infrastructure for reuse. This might sound like the perfect EAI solution, but the downside is that it's also the most invasive level of EAI. Unlike data-level and application interface-level EAI, which do not typically require changes to either the source or target applications, method-level EAI requires that you change many, if not all, enterprise applications.

In the world of EAI, you'll also hear a lot about composite applications. Composite applications are nothing more than applications bound together through a common set of services. This, of course, has been the ultimate goal of distributed objects and transactional technology for years, and is certainly the goal of method-level EAI. So, make no mistake: we are walking down the same alley here.

The enabling technology for method-level EAI is any product or standard that lets applications invoke each other's methods or access shared methods on a central server. Examples of this technology include distributed objects, based on either CORBA or COM, and the new breed of application servers. Developers implement these solutions by creating composite applications or by changing existing applications to let them share methods.

Considering the invasiveness and expense of method-level EAI, you should understand its opportunities and risks clearly when assessing its value. Sharing common business logic between many applications — and so integrating those applications — is a tremendous opportunity. However, it comes with the risk that the expense of implementing method-level EAI will outpace its value.

User Interface-Level EAI

User interface-level EAI uses the user interface as a noninvasive point of integration, as shown in Layer 3 of Figure 5.18. For instance, this approach lets you integrate existing mainframe applications through their 3270 terminal interfaces, typically when no other points of integration, such as the database or API, are available.

This process uses windows and menus to access the relevant data that must be extracted and moved to another application or data store. While it sounds like an inefficient and perhaps even ill-conceived approach — something of a stop-gap measure — there is a great deal of it going on.

Other EAI levels (as described earlier) might be more technologically appealing and efficient, but in many applications, the user is the only available mechanism for accessing logic and data. In spite of its inefficiency in getting into an application, the user interface has the advantage of not requiring any changes to the source or target application.

In the context of user-level EAI, the user interface is the EAI interface. In a process known as screen scraping, or accessing screen information through a programmatic mechanism, middleware drives a user interface (for example, 3270 user interface) to access both processes and data.

For the reasons I noted earlier, user interface-level EAI should be your last-ditch effort for accessing information from older systems. You should turn to it only when there is no well-defined application interface (for example, SAP's ALE), such as those provided by most ERP applications (albeit some ERP vendors, such as PeopleSoft, do scrape screens as their interface mechanisms), or when it doesn't make sense to leverage data-level EAI. Having said that, you needn't avoid user interface level EAI completely. In most cases, it successfully extracts information from existing applications and invokes application logic. Moreover, if you perform user interface-level EAI correctly, there's virtually no difference between the user interface and a true application interface.

The enabling technology for user-level EAI has been around for years. Most 3270 emulators, for instance, provide APIs to access screen information. Moreover, middleware vendors, such as application server and message broker vendors, provide connectors to user interfaces that let developers convert screen information directly into messages for transport to other target systems and hide the complexities from middleware users.

EAI Evolves

The future of EAI depends on whom you consult. While some see EAI evolving from plumbing and middleware to process automation, many developers believe that portals and XML will drive EAI in the near future. A new, momentum technology might also appear on the horizon.

At first, most EAI will occur through noninvasive integration projects, binding applications together at the database, method or user interface levels. I am already seeing this trend. However, the future of application integration is method-level EAI, creating a composite application to bind applications together. Clearly, the development industry likes taking baby steps into any new approach and technology, and moving from noninvasive to invasive is only natural.

However, the industry has yet to get the plumbing right to really solve the EAI problem. While many middleware vendors claim to possess the ultimate EAI solution, there's a long way to go before developers can effectively integrate all applications within an enterprise. Lingering problems include application interfaces that require new programming and maintenance, as well as the inability to provide a single solution that binds applications at both the data and method levels. Thus far, you'll have to roll your own solution using "best of breed" middleware technology. Many organizations consider that a high-risk approach.

The middleware market needs to consolidate around EAI. While message brokers are good at moving information in real time from system to system, application servers are better at providing the infrastructure for sharing common business processes. Application server and message broker vendors are working together to address this problem, but they must create a more tightly coupled solution to a typical EAI problem domain.

The next frontier is integrating EAI with process automation and workflow. This means creating a set of processes on top of your enterprise's existing processes, and binding them together to automate existing business processes. While developers have been solving workflow problems for years, integration with middleware, and now EAI, is more recent. Clearly, this is where EAI is heading — that is, if we can solve the plumbing problem.

Extensible markup language, or XML, a transplanted web standard, may simplify EAI. As the enterprise integration problem became more evident, EAI architects and developers saw the value of using XML to move information throughout an enterprise. Even more valuable, you can apply XML as a common text format to transmit information between enterprises, supporting supply chain integration efforts. As a result, many are calling XML the next EDI (electronic data interchange).

XML provides a common data exchange format, encapsulating both meta data and data. This lets different applications and databases exchange information without having to understand anything about each other. To communicate, a source system simply reformats a message, a piece of information moving from an interface, or a data record as XML-compliant text, and moves that information to any other system that can read XML. However, XML is still too new, and we've yet to understand how well it will fit with EAI. It's a text-processing standard that everyone can agree on, and not much more. The jury is still out on using XML as a common messaging and data storage standard.

However you think EAI will evolve in the future, it's important to keep your sights on the business problem it's trying to solve. Applications created over the years continue to have value — and as the cost of redeveloping those applications rises, many enterprises are making the cost-effective decision to integrate rather than rebuild critical business systems. You can count on this trend to continue as the tools and technology become more powerful and developers and enterprise architects learn more about them.

Chapter 6

Best Practices for the Analysis and Design Workflow

The Analysis and Design workflow is the heart of the Elaboration phase, where you focus on the development of the architecture for your project. Your architectural efforts in this workflow are driven by your organization/enterprise-level architectural efforts (a key aspect of the Infrastructure Management workflow of Chapter 5) — focusing on how your system will work from end-to-end all the way from the user interface to your persistent storage approach. With the organization/enterprise-level domain and technical architectures as your base, you will create a software architecture document (SAD) describing your architectural design based on the information contained in your requirements model. You will then build a technical prototype to prove that your architecture works. (This is the focus of the Implementation workflow during the Elaboration phase, covered in detail in vol. 3: *The Unified Process Construction Phase*.) Next, you will validate that it works as part of the Test workflow (Chapter 7). You may find that your organization/enterprise-level architecture will be sufficient for your project team and that all you need to do is add new business functionality to it. You may also find that your project-level architectural efforts will provide valuable feedback that your enterprise architecture team can use to improve your organization/enterprise-level architectural models.

"Build the bones, the skin (the UI), and just enough muscle to make it work. Construction is all about building the rest of the muscle."
— Ivar Jacobson

There are several reasons why you want to build a technical prototype (also called an *end-to-end* or *proof-of-concept prototype*) of your system. First, this is an important risk-reduction technique; by building a prototype, you show that the technologies you have chosen all work together thereby eliminating any "nasty integration surprises" later in your project. Second, you prove to everyone in your organization that your approach is sound. This in incredibly important to your project politically because it eliminates this avenue of attack from your project's potential detractors. Third, have a working architectural/technical prototype in place helps to build a common technical vision for your team to work towards because everyone can refer to the prototype to see how things work. This is yet another avenue for team building on for your project.

Crawl before you walk.

An important best practice for architectural efforts, and for modeling in general, is to have one person who is ultimately responsible for the architecture and corresponding design model. This role is often referred to as *Chief Architect* (particularly for your organization as a whole) or *Project Architect* for a single project. This is a role that is often taken on by your project/team lead. Architecture by committee, an approach where several people are all equally responsible for the architecture, is almost always a recipe for disaster. This approach typically results in a deadlocked architecture team comprised of several political factions, each of which has its own version of the architecture. Another common result is an architecture based on compromises — although each faction got a portion of what they wanted, nobody got everything and few people, if any, like the complete architecture. Architectural committees are often formed on the belief that consensus will result in a better solution, and sometimes it actually happens. A more effective approach is to have a benevolent dictator for a chief architect that seeks input from a variety of sources, elicits discussion of important issues, but then leads the group to a decision (occasionally a decision that needs to be dictated).

Too many cooks spoil the brew.
Too many architects spoil the system.

There are several key issues to address for the Analysis and Design workflow:
- Architecture-centric modeling practices
- Separation of concerns
- Distributed architectures
- Component-based architectures
- Framework-based architectures
- Architecting the reuse of legacy software

- Interface design
- Design patterns

6.1   Architecture-Centric Modeling Practices

In section 6.7.1 "Re: Architecture" (*Software Development*, January 1996), Larry Constantine discusses the importance of architecture to the success of your project. He focuses on iterative development — one of the major philosophies of the Unified Process — and argues that your architectural efforts will be quickly forgone in favor of the development of a flashy prototype if you do not manage your project accordingly. Constantine's message forms the basis for the goals for Analysis and Design workflow within the Elaboration phase: to model the architecture for the system that your project team is building. User interface and technical prototyping are important activities that will occur during the Elaboration phase; in fact, they are often necessary activities, but they are not the only activities. You want to set and prove your architecture early in the lifecycle of your project and then evolve it appropriately throughout the lifecycle as needed. This effort starts during the Elaboration phase and continues on into the Construction phase. In this article, Constantine provides critical insight into the best practices for architecture-based development practices.

6.2   Separation Of Concerns

One of the aspects of the enhanced lifecycle of the Unified Process that developers find confusing at first is that it contains five modeling-oriented workflows — Requirements, Infrastructure Management, Business Modeling, Analysis & Design, and Deployment — that have significant modeling aspects to them. Why not just one workflow called modeling? The reason is simple: separation of concerns. Each workflow focuses on a different and unique aspect of modeling, aspects that are related to one another yet are still separate at the same time.

Similarly, within the models themselves, you want to separate several key concerns. Experience shows that by separating your presentation logic from your business logic (which, in turn, is separated from your data logic), the resulting software is significantly more robust. Separating out these concerns results in software that is built using what is called a *layered architecture*. Layered architectures, such as the one presented in Figure 6.1, result in software that is easier to develop, easier to test, and easier to extend.

Separate modeling concerns by layering your architecture.

Figure 6.1   Layering your software architecture.

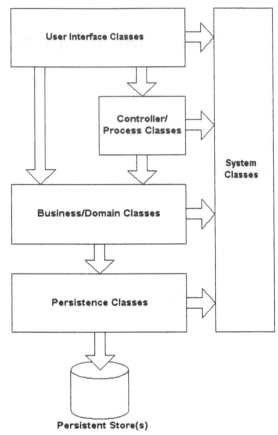

The basic idea is that you organize your software into five different types of classes:

1. User interface classes   encapsulate the screens/pages and reports that make up the user interface for your system.

2. Business/domain classes   [also known as *entity beans* in the Enterprise JavaBeans (EJB) world] implement the fundamental domain types within your application; for example, the Account and Customer classes within a bank.

3. Process classes   (such as Interest Applicator within a bank) implement complex business logic that pertains to several classes and are similar to EJB's session beans.

4. Persistence classes   encapsulate access to your persistent stores (relational databases, flat files, object bases, and so on).

5. System classes   encapsulate operating system features such as your approach to interprocess communication (IPC) and to security. This is conceptually similar to Java's "write once, run anywhere" approach.

A layered architecture is important because it makes your system more robust by promoting portability and reusabilty throughout the systems your organization develops. It also provides insight to project managers regarding how to organize your teams; each layer requires a different set of skills, therefore, you are likely to need a different personnel mix to work on each layer. For example, user interface design skills are critical for people working on the user interface layer but not so critical for people working on the system layer, where technical modeling skills are the key to success. Understanding the implications of logical architecture is a key best practice for software professionals — implications that are discussed in detail in this article.

Layered architectures provide insight into team organization.

6.3   Distributed Architectures

Organizations are becoming more and more distributed — people work from home, companies have branch offices across the globe, and "virtual organizations" are created without a brick-and-mortar office — and distributed software is often the only way to support these situations effectively. Furthermore, the scalability needs of Internet development — where you often have no way of knowing how many users will use your software, how often they will use it, and sometimes even how they will use it — require distributed architectures. In section 6.7.2 "Distributed Object Design" (*Software Development*, June 1999) I present a methodology for analyzing an object-oriented design and determining how it should be distributed. This methodology is the evolution of the design techniques presented in *Designing Object-Oriented Software* (Wirfs-Brock, Wilkerson, & Wiener, 1990). It is a multi-step process based on analyzing the message flow between the classes of your system, taking advantage of the layered class-type architecture to determine how individual classes should be distributed. The article provides a detailed look at bottom-up modeling, moving from a detailed object model comprised of both a static class model and a dynamic model (comprised of sequence diagrams and/or collaboration diagrams) to a high-level component model. This is a different approach than the one promoted by infrastructure modeling which is described in detail in Chapter 5. The technique is appropriate for traditional component architectures, discussed next, based on CORBA and Microsoft DCOM technologies where classes are distributed to specific processes in a peer-to-peer (n-tier) architecture. This differs from active object environments where individual objects move, either by being told or by determining the need themselves, to various processes as needed.

Layering enables distributed architecture.

6.4   Component and Framework-Based Architectures

Clemens Szyperski, author of the book *Component Software: Beyond Object-Oriented Programming* (ACM Press, 1998), discusses the differences between objects and components in section 6.7.3 "Components and Objects Together" (*Software Development*, May 1999). The reality is that components may be developed using either procedural-based technologies such as COBOL or using object-oriented technologies such as C++ and Java. In short, objects and components complement one another. They don't compete against each other. Szyperski explores what components are, and what they aren't, and explains how they are both similar and different from objects. He also discusses the importance of interfaces, a topic that Peter Coad and Mark Mayfield explore later in this chapter in section 6.7.5 "Designing With Interfaces For Java".

Components and objects are complementary, not competitive, technologies.

6.5 Architecting the Reuse of Legacy Software

Very few project teams have the benefit of being able to start from scratch, and in fact, the cold-hard truth of the matter is that almost every team needs to take into account existing legacy applications when they are developing new software. Although Enterprise Application Integration (EAI) — the integration of key legacy applications within your organization — is actually an aspect of the Infrastructure Management workflow (Chapter 5), it is important to note that sometimes legacy integration issues fall within the scope of a single project. In section 6.7.4 "Building The Legacy Link" (*Software Development*, November 1996) Phil Hargett presents sage advice for integrating legacy applications into your new architecture/design. Hargett shares lessons that he has learned on actual projects, discussing legacy-integration issues such as ensuring protocol independence, program abstraction, exception reporting and recovery, debugging support, performance monitoring, and security characteristics. He goes on to share a collection of legacy-integration best practices, including the use of a layered architecture and the management of different approaches by developers, to solve basic integration issues. Integrating the old with the new is the reality for modern developers. A reality that proves incredibly complex in practice and that is often ignored for the simpler issues of "green-field" development practices. Phil Hargett is one of the few writers willing to tackle this complex issue head on.

Legacy integration is a reality of modern development — one that is often ignored in favor of the simpler issues surrounding brand-new "green-field" development.

6.6 Interface Design and Design Patterns

It isn't enough to simply understand the technology that you are working with or the software process that you are following. You must also understand the nuances of your task at hand, in this case modeling. In section 6.7.5 "Designing With Interfaces For Java" (*Software Development*, April 1997) Peter Coad and Mark Mayfield share such wisdom regarding the appropriate use of Java interfaces. Excerpted out of their book *Java Design* (1997), this article describes the appropriate use of Java interfaces (a defined collection of operation signatures that a class must implement) to improve the consistency and robustness of your designs. Although languages such as C++, Smalltalk, and Visual Basic do not natively support interfaces as Java does, developers using these languages can still benefit from the modeling concepts presented in this article. Interfaces are a fundamental concept supported by the Unified

Modeling Language (UML), and any modeler wishing to be proficient with the UML will want to read this article.

The appropriate use of interfaces increases the robustness
of your designs.

In addition to interfaces, patterns are another way to improve the quality of your work. In section 6.7.6 "An Introduction to Patterns" (*Software Development,* July 1998) I present an overview of what patterns are: proven and reusable solutions to common problems given a defined context. More importantly, I introduce you to different kinds of patterns — analysis patterns, design patterns, process patterns, and organizational patterns — and present ten sample patterns from a wide variety of sources. You saw in Chapter 5 that pattern reuse (the reuse of other people's approaches to solving common problems) is one of the most productive forms of reuse your organization can achieve. This article provides an excellent starting point from which to take advantage of patterns. Also, you may want to visit the Process Patterns Resource Page, `http://www.ambysoft.com/processPatternsPage.html`, which I maintain and update on a regular basis. It contains a large collection of links and references to pattern and process resources.

Patterns provide the basis for one of the most productive forms of
reuse — the reuse of someone else's thought process.

6.7   The Articles

6.7.1 "Re: Architecture" by Larry Constantine, January 1996
6.7.2 "Distributed Object Design" by Scott W. Ambler, June 1999
6.7.3 "Components and Objects Together" by Clemens Szyperski, May 1999
6.7.4 "Building The Legacy Link" by Phil Hargett, November 1996
6.7.5 "Designing With Interfaces For Java" by Peter Coad and Mark Mayfield, April 1997
6.7.6 "An Introduction to Patterns" by Scott W. Ambler, July 1998

6.7.1 "Re: Architecture"

by Larry Constantine, January 1996

In these days of iterative and/or rapid application development (RAD), we cannot forget the importance of thinking about a project's structure.

What ever happened to software architecture? Looking at a typical sample of in-house enterprise applications and commercial off-the-shelf software packages, it's often hard to find much evidence of underlying organization. Architecture, whether in the organization of the internal functionality or in the structure of the user interface, is often among the first victims felled by today's time-boxed software projects, short release cycles, and rapid application development. There seems to be no time to think through the consequences of architectural decisions. Often there is barely time to think. Full stop. Systems are thrown together as fast as features can be thought up, with little attention to overall organization. Developers find themselves looking ahead only as far as the next line of code or the next feature.

New technology has not been of much help. Visual development tools ("Shapes to Come," Peopleware, May 1995) open up new routes to the rapid evolution of sophisticated systems, but the pace of visual development and the ease with which working applications can be constructed can also contribute to a paucity of planning. The tools may even encourage a style of development where small pieces of code that mix business logic with bits of interface programming and underlying functionality get hung onto the back of visual components, the whole interconnected by a spiderweb of message passing and event threads that nobody understands completely. The result is today's version of the classic spaghetti code, with everything connected to everything else. When every change propagates unpredictably through the web of inter-connected code, the potential for continuous software evolution shrivels.

Second Chance

Add iterative refinement to rapid prototyping and the last vestiges of architecture are likely to sink into the software swamp. That's unfortunate, because iterative prototyping is a powerful approach to delivering more usable software in less time. Prototyping allows you to deliver real capability early or to try varied approaches without a full commitment. A prototype puts something in front of users in order to get feedback based on real use. Even for relatively modest systems, it is all but impossible to get everything in the user interface right on the first try, no matter how much thought and effort you put into the design. Prototyping and iterative refinement offer a second chance — and a third and a fourth — to get user inter-faces right. With each iteration, the interface and internals are refined and enhanced, delivering more and delivering it more effectively.

Unfortunately, the structure of the first prototype, which may have been fine for proving the concept, stays around to shape the basic architecture of the evolving system. Round and round you go, and with each iteration, the system grows. New layers of features are pasted on and functional enhancements are squeezed in until the basic organization that seemed so reasonable when the system was small begins to fall apart.

It would be tempting to suggest that we should simply return to some fabled days of functional decomposition when disciplined developers spent the time with a CASE tool to work out a sound and systematic architecture for the entire system before writing the first line of code. Unfortunately, by the time the traditional approaches deliver a system architecture, a RAD team equipped with rapid visual development tools will have already shipped the working system. What we need is a way to incorporate architectural considerations into a radically accelerated development life cycle. We'd like to gain something in sounder structure without slowing down the process much.

We may need to rethink the place of software architecture in the development process. Normally, we think of architecture as something that precedes construction, but just as code can be rewritten, it can also be rearchitected. One large bank, for example, was forced to fall back on its legacy systems after an ambitious and over-extended project to build a new enterprise-wide information system had to be abandoned. The old system, an accretion of generations of COBOL kludges, had a well-earned reputation as a brittle mess that broke whenever the smallest change or correction was introduced. The bank management despaired of being able to intro-duce the varied new financial products and services needed to compete in the dynamic banking industry of today. But all was not lost. While everyone else had been caught up in the frenzy of new system development, one dedicated maintenance programmer had been quietly rebuilding parts of the legacy sys-tem, cleaning up the code section by small section and restructuring the architecture in the process. Important subsystems had been reorganized sufficiently to support continued evolution.

Renewal

The experience of this bank points to the basis for a radical reorganization of rapid iterative development processes. To make iterative refinement from prototypes work over the longer term and for larger systems, you have to keep going back to the architecture to improve it. We might call this iterative architectural refinement, or "iterative rearchitecting" if you prefer to verbify your nouns. On each successive round of design and construction, the overall structure of the program is reexamined to identify how it could be improved to support newly incorporated features better.

Refining the architecture might entail reorganizing data structures, partitioning the system into more or different subsystems, or replacing work-arounds or make-do algorithms with better code. More often than not, it will mean redesigning and rewriting some code that already worked. Although this adds to the effort in the next development iteration, it makes the system more robust in terms of further refinement, reducing the cost of future iterations. These architectural reviews are particularly good for examining the object class structure, identifying potential or needed reusable components, and looking for missed opportunities for reuse.

This is concentric development, a rationalized model for rapid prototyping with iterative refinement that starts with core functions to provide a basic set of capabilities to the user. These core capabilities are identified from a selected subset of the use cases or abstract scenarios the finished system must support ("Essentially Speaking," Peopleware, Nov. 1994). Starting with an essential core of complete use cases assures that the user has support for entire tasks, not just fragments of functionality. The system is then built in concentric layers of embellishments and enhancements. As each new layer is started, an architectural review identifies refinements or changes to the software architecture that will improve the robustness of the system in support of immediate and future enhancements. These architectural changes are added into the project workload for the current cycle of concentric development.

One company in Australia has been using a variant of this approach for four years with great success. They make three releases per year of major software. Some of the "slack time" that inevitably comes in the ups and downs of periodic releases is given over to architectural review and planning, so that the architecture is updated almost continuously.

Architectural planning can pay off big time, even under the tightest dead-lines. There is a footnote to the saga of the earnest team from Ernst & Young who took the time to architect their sys-tem in a one-day programming face-off ("Under Pressure," Peopleware, Oct. 1995). Craig Bright, Doug Smith, and Robin Taylor were joined by James Thorpe for the recent Software Challenge at ITWorld in Brisbane. They returned with a new tool (Borland Del-phi) and a revised process for what the competitors were calling "frantic application development." This time they had version control and backup as well as quick but careful analysis and architecture. They won.

6.7.2 "Distributed Object Design"

by Scott W. Ambler, June 1999

Here are nine steps to help you approach distributed object design and build robust, distributed software for mission-critical applications.

CORBA. COM+. Enterprise JavaBeans (EJB). Three leading-edge technologies based on the concept of distributing objects, often in component form, across a multi-node target environment. There is a seemingly endless supply of magazine articles and books telling you how wonderful these technologies are, telling you about their inner secrets, telling you why one is better than the other. But, how do you actually approach the design process for a distributed environment?

This article presents a process, based on the techniques presented in *Designing Object-Oriented Software* (Prentice Hall, 1990) by Rebecca Wirfs-Brock, Brian Wilkerson, and Lauren Wiener, for refactoring a traditional object design to deploy it to a distributed object environment. This process is appropriate for organizations that take a peer-to-peer approach to networking, often referred to as n-tier client/server, but don't use mobile agent technology. Although you can apply this approach to two- and three-tier client/server environments, you'll quickly discover that it's overkill in those environments. Mobile agent technology focuses on different design issues than statically distributed objects (issues such as security, persistence of mobile objects, and versioning, to name a few) that are beyond the scope of this article.

I will also not address the differences between the current implementation technologies (COM+, CORBA, EJB, and so on) for distributed object development. My experience is that although the underlying technologies change over time, the design process stays the same. Although some of the terminology has changed since the Wirfs-Brock gang first wrote their book, the fundamentals are still the same. I expect that when today's implementation technologies go the way of the dodo, as have the vast majority of implementation technologies before them, this approach will still be relevant.

The Process of Distributing Your Object-Oriented Design

The basic idea behind this process is to identify domain components, large-scale components that encapsulate collections of business/domain classes, and then to deploy those components intelligently within your hardware environment. I use a bank, a common example, throughout this article to help illustrate this process. Figure 6.2 presents a high-level overview of the bank's class diagram, using the Unified Modeling Language (UML) notation, which shows the main classes and their interrelationships for a simple teller application. There are nine steps, typically performed iteratively, to the process of distributing your object design.

1. Handle non-business/domain classes
2. Define class contracts
3. Simplify inheritance and aggregation hierarchies
4. Identify domain components
5. Define domain-component contracts

6. Simplify interfaces
7. Map to physical hardware design
8. Add implementation details
9. Distribute the user interface.

Step 1: Handle non-business/domain and process classes.

Figure 6.3 depicts a common layering of an application's classes, indicating that there are five types of classes you will build your applications from. User interface classes, such as the Screen hierarchy in Figure 6.2 encapsulate the screens and reports that make up your system's user interface. Business/domain classes, also known as entity classes, implement the fundamental domain types within your application; for example, the Account and Customer classes in Figure 6.2. Process classes, such as Interest Applicator in Figure 6.2, implement complex business logic that pertains to several classes. Persistence classes encapsulate access to your persistent stores (relational databases, flat files, object bases), and system classes encapsulate operating system features such as your approach to interprocess communication. Persistence and system classes, not shown in Figure 6.2 for the sake of simplicity, are important because they make your system more robust by promoting portability and reusability throughout the systems that your organization develops.

Figure 6.2   Initial class model for a bank.

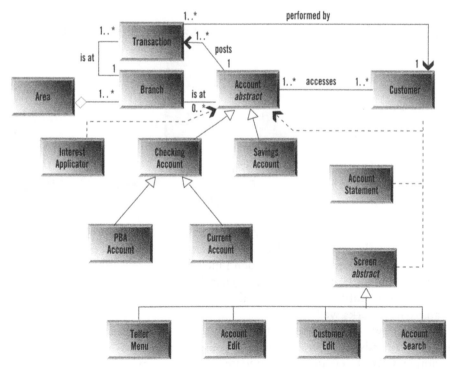

Figure 6.3 **Layering within class models.**

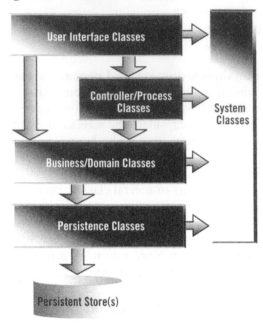

Figure 6.4 **Simplified collaboration diagram for a bank.**

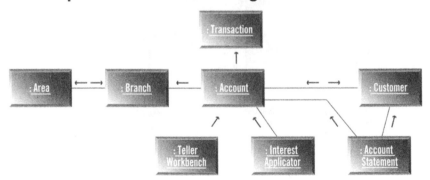

The easiest way to identify subsystems is to deal with user interface classes first, because they are generally implemented on client machines (such as personal computers, HTML browsers, and Java terminals). Although I include the Screen class hierarchy in Figure 6.4, by relabeling it the Teller Workbench, I generally choose to leave the deployment of the user interface classes until later to reduce the complexity of the problem, and in my experience, this approach works well.

There are two basic strategies for handling persistence classes, depending on your general approach to persistence. First, systems that use "data classes" to encapsulate hard-coded SQL for simple CRUD (create, read, update, delete) behavior, such as ActiveX data objects (ADOs), should deploy their data classes along with their corresponding business classes. For example, this persistence strategy would create an AccountData class that corresponds to Account, a CustomerData class that corresponds to Customer, and so on. Wherever you choose to deploy Account, you will also automatically deploy AccountData. Second, for systems that take a more robust approach to persistence, perhaps using a persistence framework or persistence layer, will typically deploy this layer on the same node as the persistent store. Organizations using several persistent stores will likely deploy the layer on its own node, which, in turn, accesses the persistent stores. For example, if your organization uses a single instance of Oracle, you will likely put your persistence layer code on the same server as Oracle. An organization that has an Oracle database, a couple of Sybase data marts, and a Versant object base will likely deploy its persistence layer or framework on its own server to provide a single source for persisting objects.

You must think hard about how you intend to deploy system classes because they comprise many critical behaviors. Obviously, you must deploy your interprocess communication classes, components, and frameworks to all nodes within your network. Other system classes, such as your security classes, will likely form a large-scale component that you may choose to deploy on its own node. You can also apply some of the following techniques for refactoring your business/domain classes and system classes to identify potential system components, such as a security manager component.

Step 2: Define class contracts.
A contract is any service or behavior of an object that other objects request. In other words, it is an operation that directly responds to a message from other classes. The best way to think about it is that contracts define the external interface, also known as the public interface, of a class. For example, the contracts of the Account class within a banking system likely include operations such as withdraw(), deposit(), open(), and close(). You can ignore operations that aren't class contracts because they don't contribute to communication between distributed classes, simplifying your problem dramatically.

Step 3: Simplify hierarchies.
For identifying servers, you can often simplify inheritance and aggregation hierarchies. For inheritance hierarchies, a rule of thumb is that if a subclass doesn't add a new contract, then you can ignore it. In general, you can consider a class hierarchy as a single class. In Figure 6.4, you see that the Account class hierarchy has been simplified as well as the Screen class hierarchy (now called Teller Workbench). For aggregation hierarchies, you can ignore any "part classes" that are not associated with other classes outside of the aggregation hierarchy. In Figure 6.4, I couldn't collapse the area/branch aggregation because account and transaction objects have associations to branch objects. Collapsing aggregation and inheritance hierarchies lets you simplify your model, so it's easier to analyze when you define subsystems.

Figure 6.5   Analyzing the collaboration diagram.

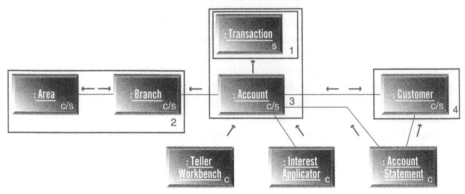

Step 4: Identify potential domain components.

A domain component is a set of classes that collaborate among themselves to support a cohesive set of contracts that you can consider blackboxes. The basic idea is that classes, and even other domain components, can send messages to domain components to either request information or request that an action be performed. On the outside, domain components appear simple, like any other type of object. On the inside, they are often complex because they encapsulate the behavior of several classes.

You should organize your design into several components in such a way as to reduce the amount of information flowing between them. Any information passed between components, either in the form of messages or the objects that are returned as the result of a message send, represents potential traffic on your network. Because you want to minimize network traffic to reduce your application's response time, you want to design your domain components so that most of the information flow occurs within the components and not between them.

To determine whether or not a class belongs in a domain component, you need to analyze its collaborations to determine its distribution type. A server class receives messages but doesn't send them, a client class sends messages but doesn't receive them, and a client/server class sends and receives messages. In Figure 6.5, each object's distribution type is indicated with the label "c," "s," or "c/s."

Once you identify the distribution type of each class, you can start identifying potential domain components. One rule of thumb is that server classes belong in a domain component and will often form their own domain components because they are the "last stop" for messages flowing within an application. The Transaction class in Figure 6.4 is a server class, but is a potential domain component in Figure 6.5. The volume of transactions, potentially millions per day, would probably motivate you to deploy Transaction to its own server (or set of servers).

If you have a domain component that is a server to only one other domain component, you may decide to combine the two components. Or, you can connect the two machines on which the domain components reside via a high-speed private link, eliminating the need to put the one machine on your regular network. In Figure 6.5, you see an example of this between Transaction and Account, indicating that you may want to go with one single domain component (indicated as box number 3) instead of two separate components. You want to consider other issues such as security (do you want to hide the existence of Transaction?), performance (will having `Transaction` on its own machine or machines prove to work faster?), and scalability (will you need to support a quickly growing number of transactions in the future?) when choosing between these two alternatives.

A third rule of thumb is that client classes do not belong in a domain component because they only generate but don't receive messages, whereas the domain component's purpose is to respond to messages. Therefore, client classes have nothing to add to the functionality a component offers. In Figure 6.5, `Interest Applicator` is a pure client class that was not assigned to a potential domain component.

A fourth rule, when two classes collaborate frequently, especially two large objects (either passed as parameters or received as return values), they should be in the same domain component to reduce the network traffic between them. Basically, highly coupled classes belong together.

A related heuristic is that client/server classes belong in a domain component, but you may have to choose which domain component they belong to. This is where you must consider additional issues, such as the information flow going into and out of the class and how the cohesiveness (how well the parts fit together) of each component would be affected by adding a new class. A fundamental design precept is that a class should be part of a domain component only if it exists to fulfill the goals of that domain component.

Finally, you should also refine your assignment of classes to domain components based on the impact of potential changes to them. One way to do this is to develop change cases ("Architecting for Change," Thinking Objectively, May 1999), which describe likely changes to the requirements that your system implements. If you know that some classes might change, then you probably want to implement change cases in one or two components to limit the scope of the changes when they occur. For example, knowing that your organization structure is likely to change over time should strengthen your resolve to encapsulate Area and Branch in a single component.

Step 5: Define domain-component contracts.

Domain-component contracts are the collection of class contracts that are accessed by classes outside of the domain component (not including class contracts that are only used by classes within the subsystem). For example, the contracts for subsystem 3 (`Account-Transaction`) would be the combination of the class contracts from the Account class hierarchy. Because transaction objects collaborate with only account objects, you can ignore the class contracts of `Transaction`. The collection of domain-component contracts form a domain component's public interface. This step is important because it helps to simplify your design, so users only need to understand each component's public interface to learn how to use it, and to decrease the coupling within your design.

You should consider several rules when defining domain-component contracts. When all of a server class's contracts are included in the contracts its domain component provides, you should consider making the server class its own domain component, external to the current component. The implication is that when the outside world needs the server class's entire public interface, encapsulating it within another domain component isn't going to buy you anything. On the same note, if none of a server class's contracts are included in the subsystem contract (for example, these contracts are only accessed by classes internal to the subsystem), you should then define the server class as a subsystem internal to the present subsystem. This heuristic would motivate you to consider including `Transaction` in the `Account` component;

however, for performance and scalability reasons mentioned previously, you might go with the component model presented in Figure 6.6, which separates them.

Figure 6.6   A component diagram for a bank.

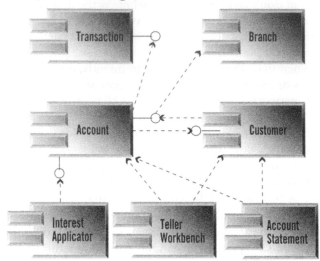

The contracts of a subsystem should be cohesive. If they don't fit well together, then you probably have multiple subsystems. If your initial contracts for the Account component include `openAccount()`, `makeDeposit()`, `makeWithdrawal()`, and `orderLunchForExecutives()`, you likely have a problem because the lunch-ordering contract doesn't fit well with the other ones.

Step 6: Simplify interfaces.
Depending on your initial design, you may be able to collapse several contracts into one to reduce the number of different message types sent to a domain component. By reducing the number of contracts for a component, you simplify its interface, making it easier to understand and hopefully use. For example, to reduce the number of contracts for `Account`, you could collapse `makeWithdraw()` and `makeDeposit()` into a single `changeBalance()` contract whose parameter is either a positive or negative amount.

Grady Booch, Ivar Jacobson, and James Rumbaugh, designers of the UML, suggest you "manage the seams in your system" (*The Unified Modeling Language User Guide*, Addison-Wesley, 1999) by defining the major interfaces for your components. They mean you should define the component interfaces early in your design, so you can work on the internals of each component without worrying about how it will affect other components — you just have to keep the defined interface stable.

Booch et al. also suggest that components should depend on the interfaces of other components and not on the component itself. In Figure 6.6, dependencies are drawn from components to the interfaces of other components. Also, notice that the `Account` component offers two different interfaces, one for the `Interest Applicator` component (likely a batch job) and one for the `Teller Workbench` and `Account Statement` components. This approach lets you restrict access to a component, making your system more secure and robust.

Figure 6.7   A deployment diagram for a bank.

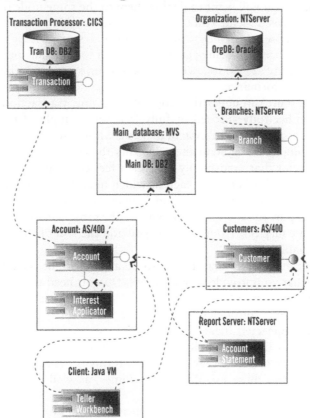

Step 7: Map to physical hardware design.

Once you identify the domain components that will comprise your system, you can map them to your physical hardware design. You will do this using the UML's deployment model, which shows your hardware nodes, the middleware used to connect them, and the components that will be deployed to each node. Figure 6.7 shows a deployment diagram for one potential configuration for this application.

Mapping domain components to hardware nodes is ideally an iterative process — you identify some candidate components, then identify some candidate hardware nodes, then you map the components to the hardware nodes, and depending on how well this works you modify either your components or your approach to the hardware configuration. You typically must map your domain components to an existing hardware configuration. You may have the option to make some changes, such as upgrading to existing machinery, but you are mostly forced to make do with what you have. In this situation, you must modify the part of your system that you still control — the design of your software.

You must consider many hard-core technical issues at this point. I discussed network performance earlier, particularly, the need to minimize network traffic. This issue now becomes critical. When you are mapping components to hardware nodes, you have partial control over what will be transmitted across your network — if two components have significant traffic between them, you will want to deploy them to the same node or at least to nodes with good connections between them. You must also consider which components can run in parallel, and ideally put those components on different nodes or at least on nodes with multiple processors. Parallel processing brings us to the issue of load balancing, spreading the processing across several hardware nodes, which makes your system configuration more complex but provides better performance.

Some behaviors must run under the scope of an atomic transaction, typically a multi-part action that must either fully succeed or fully fail. An example of this is transferring funds between two bank accounts — the system makes a withdrawal from one account and posts a transaction recording it, the system then deposits those funds into the target account and posts a transaction for that, and finally charges a fee for performing the transfer. There are five separate actions to this behavior, each of which must be performed successfully or not at all. In this case, all instances of Account and Transaction objects are affected, indicating that you may want to deploy these two components to the same hardware node. By having Transaction on a separate machine from Account, you add the risk of that connection or hardware node being unavailable.

Redundancy within your system may also be an important issue for you, improving the quality of service that you provide to your users. Do you need to deploy a single component to several nodes to support failover within your system? Failover means if one node fails, the other node with the same component can take over while the first node is being fixed. Do you have to support disconnected users? Disconnected users work with their application offline, then connect at a later time and replicate their work to the system, and then download any updates made since the last time they connected. If this is the case, then you must deploy the requisite components to these users' client machines, as well as to your network's internal nodes.

At this point, you will want to finalize where you deploy your persistence and system classes, as I indicated in step one. You should deploy persistence classes close to the persistent stores — relational databases, flat files, object databases — that they interact with, either on the same nodes or on the nodes of the domain components that need to access them directly. You should deploy system classes across all your system's nodes to manage basic behaviors, such as security and interprocess communication.

You can't do most of this step's work on paper — you must prototype the components, deploy them to a representative hardware configuration, and test them to see how well the configuration works. This is called technical prototyping and it is a critical technique for ensuring that your system design actually performs as required. You should begin technical prototyping as early in design as possible because if your design isn't going to work you want to discover this as soon as possible so that you can change your solution as required. Everything works on paper, but not always in practice.

Step 8: Add implementation details.
You've identified potential subsystems within your design and then simplified them. So what? You've drawn some bigger bubbles around some smaller bubbles. It's obvious you can't deploy bubbles, you must add something to your software to implement those bubbles. To implement domain components you need to define router classes, also known as facade classes (*Design Patterns* by Erich Gamma et al., Addison-Wesley Professional Computing,

1995), that accept incoming messages sent to the server and pass them to the appropriate classes. A router class encapsulates the server's functionality, implementing the operations for the domain-component contracts and providing an interface that other components can interact through.

A key implementation issue is whether or not your team will follow industry or organization-accepted standards and guidelines to develop your software. The Internet is successful because it was built by people following industry-accepted standards, such as HTTP and POP. Had individual development teams decided to invent "better" approaches, the Internet would never have gotten off the ground. The point is that you don't need to reinvent the wheel, you can often purchase or obtain for free significant portions of your work.

Step 9: Distribute the user interface.

The final step is to distribute your user interface appropriately, the basic goal is to deliver the appropriate screens and reports to the people who need them. You do this by looking at the actors in your use case model and seeing which use cases they interact with. You may decide to develop a user interface component that encapsulates the appropriate user interface classes, such as the `Teller Workbench` in Figure 6.6, for each actor. In the case of the bank, you would likely need to develop a user interface component that implements the user interface provided at an ATM machine and perhaps one available to senior management for operational reporting. The result of this strategy is that you will develop what appears to be several systems, the teller workbench application, the ATM application, and the management reporting application, by reusing the same domain components for each.

I've presented an approach to creating a distributed object design that starts with your non-distributed object design, analyzes it, and reworks it to be distributed. You could just as easily take an architectural driven approach where you first develop an enterprise use-case model, identify candidate classes, assign them to candidate domain components, and then flesh out the domain components via detailed modeling. I presented a bottom-up approach to distributed object design, but you can take a top-down, architectural approach.

An important lesson that you should learn is that distributed object design is doable, you just need to take several new design considerations into account. You can design robust, distributed software for mission-critical applications: the largest system in the world is the Internet, and it's highly distributed.

6.7.3 "Components and Objects Together"

by Clemens Szyperski, May 1999

The differences between the emerging component-based development and long-standing object-oriented development are often unclear. Find out how separate these concepts really are.

Components are on the upswing; objects have been around for sometime. It is understandable, but not helpful, to see object-oriented programming sold in new clothes by simply calling objects "components." The emerging component-based approaches and tools combine objects and components in ways that show they are really separate concepts. In this article, I will examine some key differences between objects and components to clarify these muddy waters. In particular, you'll see that approaches based on visual assembly tools really assemble objects, not components, but they create components when saving the finished assembly.

Why Components?

What is the rationale behind component software? Or rather, what is it that components should be? Traditionally, closed solutions with proprietary interfaces addressed most customers' needs. Heavyweights such as operating systems and database engines are among the few examples of components that did reach high levels of maturity. Large software systems manufacturers often configure delivered solutions by combining modules in a client-specific way. However, the interfaces between such modules tend to be proprietary, at most open to highly specialized independent software vendors (ISVs) that specifically produce further modules for such systems. In many cases, these modules are fused together during a linking step and are no longer distinguishable in deployed solutions.

Attempts to create low-level connection standards or wiring standards are either product or standard-driven. The Microsoft standards, resting on COM, have always been product-driven and are thus incremental, evolutionary, and, to a degree, legacy-laden by nature.

Standard-driven approaches usually originate in industry consortia. The prime example here is the effort of the Object Management Group (OMG). However, the OMG hasn't contributed much in the component world and is now falling back on JavaSoft's Enterprise Java-Beans standards for components, although it's attempting a CORBA Beans generalization. The EJB standard still has a long way to go; so far it is not implementation language-neutral, and bridging standards to Java external services and components are only emerging.

At first, it might surprise you that component software is largely pushed by desktop- and Internet-based solutions. On second thought, this should not surprise you at all. Component software is a complex technology to master — and viable, component-based solutions will only evolve if the benefits are clear. Traditional enterprise computing has many benefits, but these benefits all depend on enterprises willing to evolve substantially.

The separate existence and mobility of components, as witnessed by Java applets or ActiveX components, can make components look similar to objects.

In the desktop and Internet worlds, the situation is different. Centralized control over what information is processed when and where is not an option in these worlds. Instead, content (such as web pages or documents) arrives at a user's machine and needs to be processed there and then. With a rapidly exploding variety of content types — and open coding standards such as XML — monolithic applications have long reached their limits. Beyond the flexibility of component software is its capability to dynamically grow to address changing needs.

What a Component Is and Is Not

The separate existence and mobility of components, as witnessed by Java applets or ActiveX components, can make components look similar to objects. People often use the words "component" and "object" interchangeably. In addition, they use constructions such as "component object." Objects are said to be instances of classes or clones of prototype objects. Objects and components both make their services available through interfaces. Language designers add further irritation by discussing namespaces, modules, packages, and so on. I will try to unfold, explain, and justify these terms. Next, I'll browse the key terms with brief explanations, relating them to each other. Based on this, I'll then look at a refined component definition. Finally, I'll shed some light on the fine line between component-based programming and component assembly.

Terms and Concepts

Components. A component's characteristic properties are that it is a unit of independent deployment; a unit of third-party composition; and it has no persistent state.

These properties have several implications. For a component to be independently deployable, it needs to be well-separated from its environment and from other components. A component therefore encapsulates its constituent features. Also, since it is a unit of deployment, you never partially deploy a component.

If a third party needs to compose a component with other components, the component must be self-contained. (A third party is one that you cannot expect to access the construction details of all the components involved.) Also, the component needs to come with clear specifications of what it provides and what it requires. In other words, a component needs to encapsulate its implementation and interact with its environment through well-defined interfaces and platform assumptions only. It's also generally useful to minimize hard-wired dependencies in favor of externally configurable providers.

Finally, you cannot distinguish a component without any persistent state from copies of its own. (Exceptions to this rule are attributes not contributing to the component's functionality, such as serial numbers used for accounting.) Without state, a component can be loaded into and activated in a particular system — but in any given process, there will be at most one copy of a particular component. So, while it is useful to ask whether a particular component is available or not, it isn't useful to ask about the number of copies of that component. (Note that a component may simultaneously exist in different versions. However, these are not copies of a component, but rather different components related to each other by a versioning scheme.)

In many current approaches, components are heavyweights. For example, a database server could be a component. If there is only one database maintained by this class of server, then it is easy to confuse the instance with the concept. For example, you might see the database server together with the database as a component with persistent state. According to the definition described previously, this instance of the database concept is not a component. Instead, the static database server program is a component, and it supports a single instance: the database object. This separation of the immutable plan from the mutable instances is the key to avoiding massive maintenance problems. If components could be mutable, that is, have state, then no two installations of the same component would have the same properties. The differentiation of components and objects is thus fundamentally about differentiating between static properties that hold for a particular configuration and dynamic properties of any particular computational scenario. Drawing this line carefully is essential to curbing manageability, configurability, and version control problems.

Objects. The notions of instantiation, identity, and encapsulation lead to the notion of objects. In contrast to the properties characterizing components, an object's characteristic properties are that it is a unit of instantiation (it has a unique identity); it has state that can be persistent; and it encapsulates its state and behavior.

Again, several object properties follow directly. Since an object is a unit of instantiation, it cannot be partially instantiated. Since an object has individual state, it also needs a unique identity to identify the object, despite state changes, for its lifetime. Consider the apocryphal story about George Washington's axe, which had five new handles and four new axe-heads — but was still George Washington's axe. This is typical of real-life objects: nothing but their abstract identity remains stable over time.

Since objects get instantiated, you need a construction plan that describes the new object's state space, initial state, and behavior before the object can exist. Such a plan may be explicitly available and is then called a class. Alternatively, it may be implicitly available in the form of an object that already exists, that is close to the object to be created, and can be cloned. You'll call such a preexisting object a prototype object.

Whether using classes or prototype objects, the newly instantiated object needs to be set to an initial state. The initial state needs to be a valid state of the constructed object, but it may also depend on parameters specified by the client asking for the new object. The code that is required to control object creation and initialization could be a static procedure, usually called a constructor. Alternatively, it can be an object of its own, usually called an object factory, or factory for short.

Object References and Persistent Objects

The object's identity is usually captured by an object reference. Most programming languages do not explicitly support object references; language-level references hold unique references of objects (usually their addresses in memory), but there is no direct high-level support to manipulate the reference as such. (Languages like C provide low-level address manipulation facilities.) Distinguishing between an object — a triple definition of identity, state, and implementing class — and an object reference (just holding the identity) is important when considering persistence. As I'll describe later, almost all so-called persistence schemes just preserve an object's state and class, but not its absolute identity. An exception is CORBA, which defines interoperable object references (IORs) as stable entities (which are really objects). Storing an IOR makes the pure object identity persist.

Components and Objects

Typically, a component comes to life through objects and therefore would normally contain one or more classes or immutable prototype objects. In addition, it might contain a set of immutable objects that capture default initial state and other component resources. However, there is no need for a component to contain only classes or any classes at all. A component could contain traditional procedures and even have global (static) variables; or it may be realized in its entirety using a functional programming approach, an assembly language, or any other approach. Objects created in a component, or references to such objects, can become visible to the component's clients, usually other components. If only objects become visible to clients, there is no way to tell whether or not a component is purely object-oriented inside.

A component may contain multiple classes, but a class is necessarily confined to a single component; partial deployment of a class wouldn't normally make sense. Just as classes can depend on other classes (inheritance), components can depend on other components (import). The superclasses of a class do not necessarily need to reside in the same component as the class. Where a class has a superclass in another component, the inheritance relation crosses component boundaries. Whether or not inheritance across components is a good thing is the focus of heated debate. The theoretical reasoning behind this clash is interesting and close to the essence of component orientation, but it's beyond the scope of this article.

Modules

Components are rather close to modules, as introduced by modular languages in the early 1980s. The most popular modular languages are Modula-2 and Ada. In Ada, modules are called packages, but the concepts are almost identical. An important hallmark of modular approaches is the support of separate compilation, including the ability to properly type-check across module boundaries.

With the introduction of the Eiffel language, the claim was that a class is a better module. This seemed justified based on the early ideas that modules would each implement one abstract data type (ADT). After all, you can look at a class as implementing an ADT, with the additional properties of inheritance and polymorphism. However, modules can be used, and always have been used, to package multiple entities, such as ADTs or indeed classes, into one unit. Also, modules do not have a concept of instantiation, while classes do. (In module-less languages, this leads to the construction of static classes that essentially serve as simple modules.)

Recent language designs, such as Oberon, Modula-3, and Component Pascal, keep the modules and classes separate. (In Java, a package is somewhat weaker than a module and mostly serves namespace control purposes.) Also, a module can contain multiple classes. Where classes inherit from each other, they can do so across module boundaries. You can see modules as minimal components. Even modules that do not contain any classes can function as components.

Nevertheless, module concepts don't normally support one aspect of full-fledged components. There are no persistent immutable resources that come with a module, beyond what has been hardwired as constants in the code. Resources parameterize a component. Replacing these resources lets you version a component without needing to recompile; localization is an example. Modification of resources may look like a form of a mutable component state. Since components are not supposed to modify their own resources (or their code!), this distinction remains useful: resources fall into the same category as the compiled code that forms part of a component.

Component technology unavoidably leads to modular solutions. The software engineering benefits can thus justify initial investment into component technology, even if you don't foresee component markets.

It is possible to go beyond the technical level of reducing components to better modules. To do so, it is helpful to define components differently.

Component: A Definition

"A software component is a unit of composition with contractually specified interfaces and explicit context dependencies only. A software component can be deployed independently and is subject to composition by third parties." — Workshop on Component-Oriented Programming, ECOOP, 1996

This definition covers the characteristic properties of components I've discussed. It covers technical aspects such as independence, contractual interfaces, and composition, and also market-related aspects such as third parties and deployment. It is the unique property of components, not only of software components, to combine technical and market aspects. A purely technical interpretation of this view maps this component concept back to that of modules, as illustrated in the following definition: A component is a set of simultaneously deployed atomic components. An atomic component is a module plus a set of resources.

This distinction of components and atomic components caters to the fact that most atomic components are not deployed individually, although they could be. Instead, atomic components normally belong to a set of components, and a typical deployment will cover the entire set.

Atomic components are the elementary units of deployment, versioning and replacement; although it's not usually done, individual deployment is possible. A module is thus an atomic component with no separate resources. (Java packages are not modules, but the atomic units of deployment in Java are class files. A single package is compiled into many class files — one per class.)

A module is a set of classes and possibly non-object-oriented constructs, such as procedures or functions. Modules may statically require the presence of other modules in order to work. Hence, you can only deploy a module if all the modules it depends on are available. The dependency graph must be acyclic or else a group of modules in a cyclic dependency relation would always require simultaneous deployment, violating the defining property of modules.

A resource is a frozen collection of typed items. The resource concept could include code resources to subsume modules. The point here is that there are resources besides the ones generated by a compiler compiling a module or package. In a pure object approach, resources are serialized immutable objects. They're immutable because components have no persistent identity. You cannot distinguish between duplicates.

Interfaces

A component's interfaces define its access points. These points let a component's clients, usually components themselves, access the component's services. Normally, a component has multiple interfaces corresponding to different access points. Each access point may provide a different service, catering to different client needs. It's important to emphasize the interface specifications' contractual nature. Since the component and its clients are developed in mutual ignorance, the standardized contract must form a common ground for successful interaction.

What nontechnical aspects do contractual interfaces need to obey to be successful? First, keep the economy of scale in mind. Some of a component's services may be less popular than others, but if none are popular and the particular combination of offered services is not either, the component has no market. In such a case, the overhead cost of casting a particular solution into a component form may not be justified.

Notice, however, that individual adaptations of component systems can lead to developing components that have no market. In this situation, component system extensions should build on what the system provides, and the easiest way of achieving this may be to develop the extension in component form. In this case, the economic argument applies indirectly: while the extending component itself is not viable, the resulting combination with the extended component system is.

Second, you must avoid undue market fragmentation, as it threatens the viability of components. You must also minimize redundant introductions of similar interfaces. In a market economy, such a minimization is usually the result of either early standardization efforts in a market segment or the result of fierce eliminating competition. In the former case, the danger is suboptimality due to committee design, in the latter case it is suboptimality due to the nontechnical nature of market forces.

Third, to maximize the reach of an interface specification, and of components implementing this interface, you need common media to publicize and advertise interfaces and components. If nothing else, this requires a small number of widely accepted unique naming schemes. Just as ISBN (International Standard Book Number) is a worldwide and unique naming scheme to identify any published book, developers need a similar scheme to refer abstractly to interfaces by name. Like an ISBN, a component identifier is not required to carry any meaning. An ISBN consists of a country code, a publisher code, a publisher-assigned serial number, and a checking digit. While it reveals the book's publisher, it does not code the book's contents. The book title may hint at the meaning, but it's not guaranteed to be unique.

Explicit Context Dependencies

Besides specifying provided interfaces, the previous definition of components also requires components to specify their needs. That is, the definition requires specification of what the deployment environment will need to provide, so that the components can function. These needs are called context dependencies, referring to the context of composition and deployment. If there were only one software component world, it would suffice to enumerate required interfaces of other components to specify all context dependencies. For example, a mail-merge component would specify that it needs a file system interface. Note that with today's components, even this list of required interfaces is not normally available. The emphasis is usually just on provided interfaces.

In reality, several component worlds coexist, compete, and conflict with each other. At least three major worlds are now emerging, based on OMG's CORBA, Sun's Java, and Microsoft's COM. In addition, component worlds are fragmented by the various computing and networking platforms. This is not likely to change soon. Just as the market has so far tolerated a surprising multitude of operating systems, there will be room for multiple component worlds. Where multiple worlds share markets, a component's context dependencies specification must include its required interfaces and the component world (or worlds) for which it has been prepared.

While the role of components is to capture a software fragment's static nature, the role of objects is to capture the dynamic nature of the arising systems built out of components.

There will, of course, also be secondary markets for cross-component-world integration. In analogy, consider the thriving market for power-plug adapters for electrical devices. Thus, bridging solutions, such as the OMG's COM and CORBA Interworking standard, mitigate chasms.

Component Weight

Obviously, a component is most useful if it offers the right set of interfaces and has no restricting context dependencies; that is, if it can perform in all component worlds and requires no interface beyond those whose availability is guaranteed by the different component worlds. However, few components, if any, would be able to perform under such weak environmental guarantees. Technically, a component could come with all required software bundled in, but that would clearly defeat the purpose of using components in the first place. Note that part of the environmental requirements is the machine on which the component can execute. In the case of a virtual machine, such as the Java Virtual Machine, this is a straightforward part of the component world specification. On native code platforms, a mechanism such as Apple's fat binaries (which pack multiple binaries into one file), would still allow a component to run everywhere.

Instead of constructing a self-sufficient component with everything built in, a component designer may opt for maximal reuse. Although maximizing reuse has many advantages, it has one substantial disadvantage: the explosion of context dependencies. If designs of components were, after release, frozen for all time, and if all deployment environments were the same, this would not pose a problem. However, as components evolve, and different environments provide different configurations and version mixes, it becomes a showstopper to have a large number of context dependencies. To summarize: maximizing reuse minimizes use. In practice, component designers have to strive for a balance.

Component-Based Programming vs. Component Assembly

Component technology is sometimes used as a synonym for "visual assembly" of pre-fabricated components. Indeed, for relatively simple applications, wiring components is surprisingly productive — for example, JavaSoft's BeanBox lets a user connect beans visually and displays such connections as pieces of pipework: plumbing instead of programming.

It is useful to take a look behind the scenes. When wiring or plumbing components, the visual assembly tool registers event listeners with event sources. For example, if the assembly of a button and a text field should clear the text field whenever the button is pressed, then the button is the event source of the event button pressed and the text field is listening for this event. While details are of no importance here, it is clear that this assembly process is not primarily about components. The button and the text field are instances, that is, objects not components. (When adding the first object of a kind, an assembly tool may need to locate an appropriate component.)

However, there is a problem with this analysis. If the assembled objects are saved and distributed as a new component, how can this be explained? The key is to realize that it is not the graph of particular assembled objects that is saved. Instead, the saved information suffices to generate a new graph of objects that happens to have the same topology (and, to a degree, the same state) as the originally assembled graph of objects. However, the newly generated graph and the original graph will not share common objects: the object identities are all different.

You should then view the stored graph as persistent state but not as persistent objects. Therefore, what seems to be assembly at the instance rather than the class level — and is fundamentally different — is a matter of convenience. In fact, there is no difference in outcome between this approach of assembling a component out of subcomponents and a traditional programmatic implementation that hard codes the assembly. Visual assembly tools are free to not save object graphs, but to generate code that when executed creates the required objects and establishes their interconnections. The main difference is the degree of flexibility in theory. You can easily modify the saved object graph at run time of the deployed component, while the generated code would be harder to modify. This line is much finer as it may seem — the real question is whether components with self-modifying code are desirable. Usually they are not, since the resulting management problems immediately outweigh the possible advantages of flexibility.

It is interesting that persistent objects, in the precise sense of the term, are only supported in two contexts: object-oriented databases, still restricted to a small niche of the database market, and CORBA-based objects. In these approaches, object identity is preserved when storing objects. However, for the same reason, you cannot use these when you intend to save state and topology but not identity. You would need an expensive deep copy of the saved graph to effectively undo the initial effort of saving the universal identities of the involved objects.

On the other hand, neither of the two primary component approaches, COM and Java-Beans, immediately support persistent objects. Instead, they only emphasize saving the state and topology of a graph of objects. The Java terminology is "object serialization." While object graph serialization would be more precise, this is much better than the COM use of the term persistence in a context where object identity is not preserved. Indeed, saving and loading again an object graph using serialization (or COM's persistence mechanisms) is equivalent to a deep copy of the object graph. (Many systems use this equivalence to implement deep copying.)

While it might seem like a major disadvantage of these approaches compared to CORBA, note that persistent identity is a heavyweight concept that you can always add where needed. For example, COM supports a standard mechanism called monikers, objects that resolve to other objects. You can use moniker to carry a stable unique identifier (a surrogate) and the information needed to locate that particular instance. The resulting construct is about as heavyweight as the standard CORBA Object References. Java does not yet offer a standard like COM monikers, but you could add one easily.

Component Objects

Components carry instances that act at run time as prescribed by their generating component. In the simplest case, a component is a class and the carried instances are objects of that class. However, most components (whether COM or JavaBeans) will consist of many classes. A Java Bean is externally represented by a single class and thus is a single kind of object representing all possible instantiations or uses of that component. A COM component is more flexible. It can present itself to clients as an arbitrary collection of objects whose clients only see sets of unrelated interfaces. In JavaBeans or CORBA, multiple interfaces are ultimately merged into one implementing class. This prevents proper handling of important cases such as components that support multiple versions of an interface, where the exact implementation of a particular method shared by all these versions needs to depend on the version of the interface the client is using. The OMG's current CORBA Components proposal promises to fix this problem.

Mobile Components vs. Mobile Objects

Surprisingly, mobile components and objects are just as orthogonal as regular components and objects. As demonstrated by the Java applet and ActiveX approaches, it is useful to merely ship a component to a site and then start from fresh state and context at the receiving end. Likewise, it is possible to have mobile objects in an environment that isn't component-based at all. For example, Modula-3 Network Objects can travel the network, but do not carry their implementation with them. Instead, the environment expects all required code to already be available everywhere. It is also possible to support both mobile objects and mobile components. For example, a mobile agent (a mobile autonomous object) that travels the Internet to gather information should be accompanied by its supporting components. A recent example is Java Aglets (agent applets).

The Ultimate Difference

While components capture the static nature of a software fragment, objects capture its dynamic nature. Simply treating everything as dynamic can eliminate this distinction. However, it is a time-proven principle of software engineering to try and strengthen the static description of systems as much as possible. You can always superimpose dynamics where needed. Modern facilities such as meta-programming and just-in-time compilation simplify this soft treatment of the boundary between static and dynamic. Nevertheless, it's advisable to explicitly capture as many static properties of a design or architecture as possible. This is the role of components and architectures that assign components their place. The role of objects is to capture the dynamic nature of the arising systems built out of components. Component objects are objects carried by identified components. Thus, both components and objects together will enable the construction of next-generation software.

Blackbox vs. Whitebox Abstractions and Reuse

Blackbox vs. whitebox abstraction refers to the visibility of an implementation behind its interface. Ideally, a blackbox's clients don't know any details beyond the interface and its specification. For a whitebox, the interface may still enforce encapsulation and limit what clients can do (although implementation inheritance allows for substantial interference). However, the whitebox implementation is available and you can study it to better understand what the box does. (Some authors further distinguish between whiteboxes and glassboxes, where a whitebox lets you manipulate the implementation, and a glassbox merely lets you study the implementation.)

Blackbox reuse refers to reusing an implementation without relying on anything but its interface and specification. For example, typical application programming interfaces (APIs) reveal no implementation details. Building on such an API is thus blackbox reuse of the API's implementation. In contrast, whitebox reuse refers to using a software fragment, through its interfaces, while relying on the understanding you gained from studying the actual implementation. Most class libraries and application frameworks are delivered in source form and application developers study a class implementation to understand what a subclass can or must do.

There are serious problems with whitebox reuse across components, since whitebox reuse renders it unlikely that the reused software can be replaced by a new release. Such a replacement will likely break some of the reusing clients, as these depend on implementation details that may have changed in the new release.

6.7.4 "Building The Legacy Link"

by Phil Hargett, November 1996

A homegrown middleware solution requires a carefully thought-out design. See how one company achieved this using Visual Basic and Windows NT.

Many client/server applications provide highly usable GUI front ends to legacy databases through well-established, off-the-shelf middleware: ODBC, CORBA, DCE, or interfaces provided by specific database vendors such as Sybase, Oracle, or SQL Server. In some cases, the "plumbing" between the user interface and the legacy systems is barely more than a simple data stream, as was the case with a recent call center project I worked on with a Fortune 100 firm. The plumbing or middleware for this project was a slender interface to the advanced program-to-program communication (APPC) services provided by Microsoft SNA Server.

While the middleware offered little help to the programmer, my team looked at this project as an opportunity to gain the design skills necessary to produce an application based on proprietary middleware (which in this case required the programmer to do everything). Much design work in the market has centered on legacy RDBMS access, and less on raw data stream access to legacy data. Further, the chance to use newer technologies like Visual Basic, OLE Automation, and Windows NT was an added benefit.

To minimize the risk of a bad link to legacy applications, the application team followed a specific set of design principles. The design we chose to adopt was less important than the issues it addressed, and how well it handled them.

What Makes a Good Link?

Like any client/server application, the application transmitted requests from the client application code to the server program code, as shown in Figure 6.8. We termed these requests

Business Service requests. The client application requests Business Service to perform work on its behalf, such as updating a data store, performing a search, or calculating a result. A Business Service was usually just a wrapper around existing legacy code, but in some cases it was a completely new program. Each request was answered by a Business Service response, containing either the result of the operation or detailed information for reporting error conditions. A good design for this link between applications — the plumbing — handles such requests (and the associated responses) in a robust, consistent manner, as well as addresses a number of issues. The following list of issues (by no means exhaustive) applies to most mechanisms for implementing this link (CORBA, ODBC, and so on).

Figure 6.8   Business service requests.

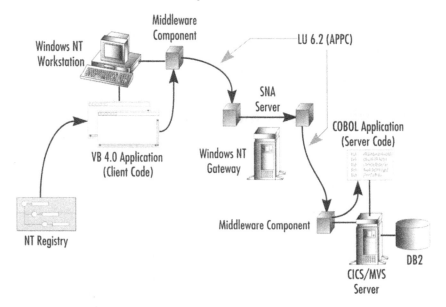

Protocol independence   Different communications protocols have distinct APIs and may create or require distinct characteristic behaviors in client applications. As much as possible, the application should insulate itself from the specific APIs or behavioral characteristics of the selected communication protocols.

Routing and configuration management   Client applications rely on named services and the ability to use a name lookup function at run time to locate the exact service to execute. How services are named, who knows those names, and where the name lookup function resides are some of the key decisions a designer must make. There are many possible designs, but maintaining the list of names should be easy, and any name lookup functions should contribute as little overhead as possible.

Pay attention to the application's degree of fan-out during the design process. The fan-out is the number of server locations any single client application must know of to perform its work. Location may be defined in terms of logical names of network sites, specialized names or nodes within a site, or even the names of particular programs within a node. You should strive to minimize fan-out, if possible. As a client/server application evolves, the location of services typically evolves as well. Any module with a high degree of fan-out may also have higher maintenance overhead.

Program abstraction   Ideally, client and server code should not be aware that certain portions of the application are running on different platforms; this quality is known as location transparency. Designers should strive to encapsulate — or limit — the functional dependencies between modules. Program abstraction will ensure location transparency and encapsulation of function. That's why remote procedure call (RPC) is an effective paradigm for client/server development: RPCs look no different than those executed locally. Other abstractions may exist: message-queuing, CORBA, or Berkeley Software Distribution Sockets, to name a few.

In a legacy environment, program abstraction has additional significance. As new systems roll into production to replace aging source legacy applications, abstracting a Business Service's function from its source system becomes essential to minimizing the impact over time as source systems change.

Exception reporting and recovery   Errors will always exist in a client/server environment. A transport failure that brings down the application is only one that might occur. The link to legacy applications should have a clean, consistent mechanism for detecting exceptions, logging information for diagnosis, and recovering from cross-platform exceptions.

Debugging support   Many times an exception occurs not because of a transport failure, but because the client and server programs are simply not speaking the same language. The ability to examine the packets transmitted and received between the client and server platforms is vital to debugging the application.

Performance monitoring support   Performance monitoring in a client/server environment is nontrivial. There may be lags on client components, server components, or network components. Support for performance monitoring (and tuning, if possible) should be considered in all aspects of the design.

Security characteristics   Conscious choices about how the application's legacy plumbing addresses security issues is important. Identifying how the application will relate to existing security packages and policies will be critical for acceptance of the application's design in a production environment. Designers should address security concerns in all environments where the application operates.

Marshaling complexity   Marshaling is when a program translates a data structure from its own platform-specific format to a structure that the network protocol can transmit without error. The receiving platform translates into its own platform-specific data structure with the same semantic meaning as that found in the sending platform's structure. You should minimize marshaling complexity as much as possible.

Because the middleware component in the call center project delivered only a small portion of the functionality required, we decided to develop a set of principles to guide the design of the application's strategy for linking to legacy applications.

Implementation

Business Service Requests traveled a long path from client to server and back again, as Figure 6.8 illustrates. A request could be almost anything: a search request, an update to server databases, or a check of a user's security profile. Each request began inside the 32-bit Visual Basic

4.0 application. As the application made a request, it first looked inside the Windows NT registry for routing information. The registry informed the application what server on the network should receive the request, what program to execute on that server, and what transaction under which the program should run. The middleware could only transmit textual data, but it performed conversions between EBCDIC and ASCII as necessary. The client application, having prepared the request for transmission and attached the configuration information, used the middleware component to send the request to the appropriate server and waited for a response.

Once the target service program received and processed the request, it returned the results to the client application. The client application remained in suspension, reawakening only when a response arrived back at the workstation.

Software Architecture

Three types of modules accounted for most of the project's source code, as shown in Figure 6.9. Form modules managed the user interface, one for each window in the application. All Business Service logic was hidden within a collection of distinct Service class modules, one specialized class for each Business Service needed by the application. The final component was a single module, the Request module, which managed cross-platform communications for the entire application.

Figure 6.9   Modules accounting for source code.

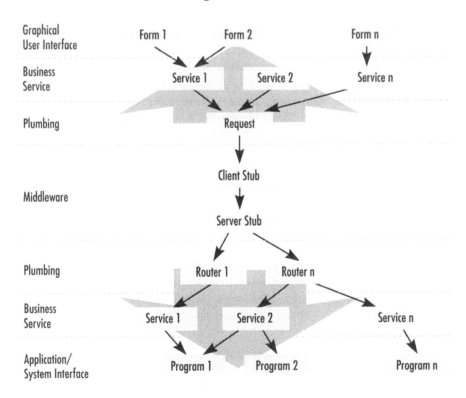

A Request object represented the lowest available interface in Visual Basic to the communications protocols necessary for communicating Business Service requests from a client program to the appropriate server program. An application only required one class of Request objects, although many instances may be created (one per instance of a Service object). A Request object had three key responsibilities:

1. To hide details of the specific protocol that it used for communications.
2. To acquire the necessary configuration information to communicate with a specific program on another platform.
3. To buffer the communication: the object held data until the client application actually started the request. After the communication was complete, it retained the results returned by the server program.

The Request object offered several functions in its interface, as shown in Table 6.1.

A Service object defined an interface to a specific Business Service. Each Service object had several duties:

• To manage the marshaling and unmarshaling of data.
• To provide mechanisms for client applications to initiate or transmit a request to the server program and check on the status of the request .
• To let client applications exchange data elements with the Service object as necessary.
• To perform its services without relying on GUI dependencies and without assuming that any particular module had invoked its methods.
• To maintain a one-to-one correspondence between Business Services implemented on the remote platform by acting as the client's sole interface to that Service.

The definition of each Service included two data structures that were equally represented in both Visual Basic and COBOL: one structure defining the input parameters for the Service, and one defining the output. The Service functions are shown in Table 6.2.

Table 6.1   Request functions.

Function	Description
Configure	Takes the name of a configuration and retrieves all of the relevant parameters from the Windows NT Registry.
Set Data	A program called this routine to pass the data being sent to the server into a Request object.
Start	Initiates the transmission and waits for a response.
Status	A program could call this routine to determine the success or failure of a transmission.
GetData	A program called this routine to retrieve the data returned from the server (including error data) from a Request.

Table 6.2   Service functions.

Function	Description
Initialize	Similar to a constructor in C++, this routine managed the acquisition and configuration of a Request object.
Marshal	An internal routine to translate the Service's input data from the PC format to the server's format for the data.
DoIt	Once a program had passed the required input parameters to the Service module, the program invoked this routine to marshal all the parameters, pass the data to the Request object, start the transmission, check the status on completion, and unmarshal the results for further use.
Unmarshal	An internal routine to translate back again the Service's result from a server format back to the PC's format.
Status	A program could call this routine to determine the success or failure of the transmission.

Modules above the level of Services, such as Forms, also had some characteristics that were important to this design:

- No Form contained code whose purpose was the retrieval of data for use on another Form.
- Forms (or other modules) neither required nor exploited any knowledge of the internal data structures of the Business Service, including both target server data structures or the data structures used by any of its Service objects.
- Forms neither required nor exploited knowledge of the other Forms' internal data structures. For example, many search screens permitted the user to select a single record for displaying further information about that record on a secondary detail screen or Form. The secondary Form was able to discover the key to that record not by examining a control on the search Form, but only through a public variable on the search Form. If the search Form maintained that interface, the detail Form would not be affected by internal changes to the search Form.

Requests and Responses

The exact contents of each request or response transmitted between platforms varied according to the Business Service involved, although each transmission had a consistent design. Table 6.3 lists some of the standard elements for different types of transmission.

Requests from Client to Server

These structures (like all structures passed between client and server) were of fixed length: the individual elements transmitted requests, and responses were of fixed length, fixed number of occurrences, and fixed position. If either a request or response structure contained an occurring structure, then there would be an additional element indicating how many occurrences contain valid data.

Table 6.3   Standard data elements for transmission.

Requests from Client to Server	Responses from Server to Client	Errors Returned from Server to Client
• Version number • Request or service code • User ID • Machine ID • Length in bytes of total request • Length in bytes of service-specific request data	• Result code (Boolean value) • Length in bytes of total response structure • Length of bytes of service-specific response data	• User ID • Machine ID • CICS task ID • Date and time • Request or service code • Module name • Text message • Several CICS-specific fields, including abend code • Several DB2-specific fields, including • SQLCODE

If a service encountered significant errors on the host, then the service might return the standard error data structure instead of its own response structure. When this occurs, standard routines would post the error in the same fashion as if code on the workstation had generated it. Client code specific to each Business Service understood the layout of the service's request and response structures, while standard client routines shared by all managed the error structure.

Evaluation

When evaluating how the application and its enabling technologies adhere to the design principles, one important technology is worthy of special attention. Windows NT provides many administration utilities to the user as part of the base installation of the operating system. These tools, such as the registry editor, the event viewer, and the performance monitor all have the important characteristic of being remotable: from a single workstation, they may be used to monitor or configure any workstation available on the wide-area network, security permitting. These remotable tools are not restricted to specialized, administrative installations of Windows NT — all installations of Windows NT include them.

Protocol independence.   The application was initially developed and tested with a 16-bit middleware component using Named Pipes and Logical Unit 6.2 (APPC) in combination, but later development and production releases used 32-bit middleware exclusively for Logical Unit 6.2. The design of the application's plumbing delivered a single API (the Request class module) that did not change as the underlying protocols evolved.

Routing and configuration management.   The architecture described here primarily localized fan-out on the server platform, but allowed for rapid alteration of fan-out at the client through the NT registry to handle special situations such as performance monitoring or debugging. Although the application invoked more than 70 host services, each workstation required only three configurations to reach host-based routers that would distribute requests to the appropriate service. The use of each machine's registry for configuration greatly simplified the process of managing routing information.

Program abstraction. The results of the application's support for abstraction are ambiguous. Many Service modules obeyed the rules regarding what they could and couldn't do; however the real test of this principle would have been the effort required to reuse the module in another application or in another part of the same application. Except as templates for creating new services, most of the services were too specialized to reuse in their entirety. Other projects did, however, borrow the entire Request module.

More could be said about why this principle was not successfully upheld within the application. What is important to remember is that well-designed plumbing is not the only key to success in a client/server application, even though poorly designed plumbing can potentially cause the most grief. Good design must prevail throughout.

Despite some stability concerns when using OLE in the early stages of development, the project could have benefited from a more rigorous object-oriented, OLE-centered design. All Business Services and the Request module made extensive use of OLE automation, as each was implemented as a separate class of object. Better use of OLE technology for packaging object components would have enabled the partitioning of work among developers and a cleaner application design.

Exception reporting and recovery. The architecture met this principle, through two key means: centralized error-handling in the Request module, and standard error-handling code residing in the common Service template copied by almost all developers when beginning to write a new service. The Windows NT event log on each workstation provided an excellent facility for capturing diagnostic information as errors occurred.

Performance monitoring support. The application provided reasonable support for performance monitoring. Within the central Request module shared by all developers, logic was added to record the time required for the host to respond to a client request. The extra logic recorded numerous parameters (such as machine name, user ID, service name, and so on) as well as the transmitted packets in their entirety. The Windows NT performance monitor enabled tracking of many operating system statistics as the application ran: the size of the system swap file, the number of pages swapped in or out each second, and the size of the working set, to name a few.

Debugging support. By encapsulating communications in an object, developers could examine each instance of the communications code to understand the state of communications within the application. Also, by encapsulating each Business Service within its own object, a particular instance of a Business Service could provide richer access to debugging information. Regardless of the service being debugged, a programmer could always evaluate the same variables and statements to determine what was happening inside the application. A drawback to the implementation of this principle was the high degree of duplicate data maintained by the application to achieve this goal.

Security characteristics. Perhaps a kinder assessment may have said the application addressed the issue well. However, the chosen approach to security actually avoided the issue entirely. The application invoked all of its Customer Information Control System (CICS) transactions using the default user ID assigned to the central routing region — a situation not often considered acceptable because it eliminates the possibility of applying the finer-grained security tools available in CICS and DB2. Many information systems departments would have thrown the design out the window had it appeared in their shops!

Marshaling complexity. The application did not meet this requirement at all. Although the developers exploited many tricks and techniques and tried numerous avenues to avoid complexity, the many parts of the application dealing with this issue remain an over-complicated mass of code that will have a high maintenance cost. A great shortcoming of Visual Basic is its inability to perform such manipulations naturally, outside of the standard mechanisms employed by OLE. C or C++ would have been kinder to the developers regarding these matters.

The client chose Visual Basic prior to the project, and the development was reasonably comfortable with the language and the design. Unfortunately, some of the most arcane code in the application was code required to marshal and unmarshal the packets of textual data transmitted and exchanged with the host services. The application used string manipulation functions extensively. The lack of type coercion or the equivalent of a C language union construct proved a hindrance to this task. In defense of Visual Basic, the project's design did not exploit Visual Basic's natural integration with OLE for this task; Visual Basic provides automatic marshaling and unmarshaling between a client program and an OLE server. Had the project written small custom stubs in C++ to handle the low-level interface to the host services (an easy task), then the advantages of OLE would have been useful.

One of the more difficult challenges is not uncommon on many client/server projects: integrating different design approaches taken by programmers specializing in different platforms. GUI programmers often do not understand the designs favored by mainframe programmers, and vice versa. Competent design skills for a specific platform do not guarantee success in client/server design. Learning how to address the issues described here and still deliver an application that's well designed on both platforms is an immense challenge. Developing common thought processes among mainframe and PC programmers is not easy, never mind adding the additional complexities of client/server architecture. However, get the basic mechanics of cross-platform communication right, and the rest of the problem can be solved by the skills of those parties best suited to each platform.

6.7.5 "Designing With Interfaces For Java"
by Peter Coad and Mark Mayfield, April 1997

Whether you use DCOM or CORBA, interfaces are the key to pluggability.

Designing with interfaces is the most significant aspect of Java-inspired design. It provides you with freedom from object connections that are hardwired to just one class of objects, as well as freedom from scenario interactions that are hardwired to just one class of objects. For systems in which flexibility, extensibility, and pluggability are key issues, Java-style interfaces are a must. Indeed, the larger the system, and the longer its potential life span, the more significant interface-centric scenario development becomes to the system's designers. In this article, we'll explore Java-style interfaces: what they are, why they are important, and two major contexts where they're helpful.

Design with Interfaces? Yes!

Interfaces are the key to pluggability, the ability to remove one component and replace it with another. Consider the electrical outlets in your home: the interface is well-defined (plug shape, receptacle shape, voltage level for each prong). You can readily unplug a toaster, plug in a coffeemaker, and continue on your merry way.

An interface is a common set of method signatures that you define once, and then use again and again in your applications. It's a listing of method signatures alone; there is no common description, no source code behind any of those method signatures.

Java expresses inheritance and polymorphism distinctly, with different syntax. C++ expresses both concepts with a single syntax; it blurs the distinction between these very different mechanisms, resulting in overly complex, overly deep class hierarchies. (You can design with interfaces either way; it's just that Java makes it easier to express that design in source code.) An interface describes a standard protocol, a standard way of interacting with objects in classes that implement the interface.

Working with interfaces requires you to specify the interface and which classes implement it. Let's begin with a simple interface, called IName. IName consists of two method signatures, the accessors getName and setName. In Java, an IName interface might look something like this:

```
public interface IName {
    String getName();
    setName(String aName); }
```

A class that implements the IName interface promises to implement the getName and setName methods in an appropriate way for that class. An example is shown in Figure 6.10. The getName method returns the name of an object. The setName method establishes a new name for an object.

The IName interface describes a standard way to interact with an object in any class that implements that interface. This means that as an object in any class, you could hold an IName object (objects within any number of classes that implement the IName interface). You could also ask an IName object for its name without knowing or caring about what class that object happens to be in.

Figure 6.10   An interface and a class.

*An interface and a class that
promises to implement that interface*

Why Interfaces?

Over the years, we've encountered a classic barrier to three things: flexibility (the ability to graciously accommodate changes in direction), extensibility (the ability to graciously accommodate add-ons), and pluggability (the ability to graciously accommodate pulling out one class of objects and inserting another with the same method signatures). Yes, this is a barrier within object-oriented design itself.

Some background is in order. All objects interact with other objects to get something done. An object can answer a question or calculate a result all by itself; yet even then, some other object does the asking. That's why scenario views are so significant, capturing time-ordered sequences of object interactions. But in object models and scenario views, an object must be within a specified class. That's a problem.

What is the element of reuse? It's not just a class; objects in the specified class are interconnected with objects in other classes. Instead, it's some number of classes, or the number of classes in a scenario, or even the number of classes in overlapping scenarios.

What's the impact, in terms of pluggability? If you want to add another class of objects, one that can be plugged-in as a substitute for an object in another class that's already in a scenario view, you are in trouble. There is no pluggability there. Instead, you must add object connections, build another scenario view, and implement source code behind it all. The problem with doing so is that each object connection and each message-send is hardwired to objects in a specific class. This impedes pluggability, extensibility, and flexibility.

Traditionally, objects in a scenario view are hardwired to each other. Yet if the "what I know" (object connections) and "who I interact with" (object interactions) are hardwired to just one class of objects, then pluggability is nonexistent. Adding a new class means adding the class itself, object connections, scenario views, and changes to other classes — in the design and in the source code.

A more flexible, extensible, and pluggable approach — one that would let you add new classes of objects with no change in object connections or message-sends — would definitely be preferrable. Fortunately, there is a partial solution. To add a new class that's a subclass of one of the classes of objects participating in a scenario, you simply add a specialization class to your object model, then add a comment to your scenario view that states the objects from the specialization class are applicable. However, if inheritance does not apply, or if you have already used it in another way (keeping in mind that Java is a single-inheritance language), then this partial solution is no solution at all.

Interfaces enhance, facilitate, and allow for flexibility, extensibility, and pluggability. They shift how people think about an object and its connections and interactions with other objects. For example, you might ask, "Is this connection (or message-send) hardwired to objects in that class; or is this a connection (or message-send) to any object that implements a certain interface?" If it's the latter, you would actually be saying, "I don't care what kind of object I am connected (or sending messages) to, just as long as it implements the interface I need." Therefore, when you need flexibility, you must specify object connections (in object models) and message-sends (in scenario views) to objects in any class that implements the interface that is needed, rather than to objects in a single class (or its subclasses).

Interfaces loosen up coupling, make parts of a design more interchangeable, and increase the likelihood of reuse — all for a very modest increase in design complexity.

With interfaces, composition plays an increasingly important role. It becomes flexible, extensible, and pluggable (composed of objects that implement an interface), rather than hardwired to just one kind of object (composed of objects in just one class). Interfaces also reduce the otherwise compelling need to jam many classes into a hierarchy with a lot of multiple inheritance. In effect, using interfaces streamlines how you use inheritance: you use interfaces to express generalization-specialization of method signatures, and you use inheritance to express generalization-specialization of implemented interfaces — along with additional attributes and methods.

Interfaces let you separate method signatures from method implementations. You can use them to separate user interface method signatures from operating system-dependent method implementations; that's exactly what Java's Java Foundation Classes (JFC) and their predecessor the Abstract Windowing Toolkit (AWT) do. You could also do the same for data management or even for problem-domain objects.

In effect, interfaces give you a way to establish object connections and message-sends to objects in any class that implements a needed interface — without hardwiring to object connections or message-sends to a specific class of objects. The larger the system is, and the longer its potential life span is, the more significant interfaces become.

Bringing Interfaces into Your Design

Factoring out every method signature into a separate interface would be overkill — you'd make your object models more complex and your scenario views too abstract. One way to bring interfaces into your design is to factor out method signatures into a proxy. A proxy is one that acts as a substitute. Consider person and passenger in a flight-reservation system. A person object might have accessors like getName. A passenger object could act as a proxy. When someone asks that passenger object for its name, it would (in turn, behind the scenes) ask its corresponding person object for its name, and then return the result to the sender.

Why bother with a proxy? Consider the design before you let one object act as a proxy for another. The non-proxy scenario looks like Figure 6.11.

Using a proxy, the scenario is simpler, as shown in Figure 6.12. Hence, with a proxy, scenario views become simpler; the details about whomever is being represented by the proxy is shielded from view. The corresponding object model looks like what is shown in Figure 6.13.

Figure 6.11   The non-proxy scenario.

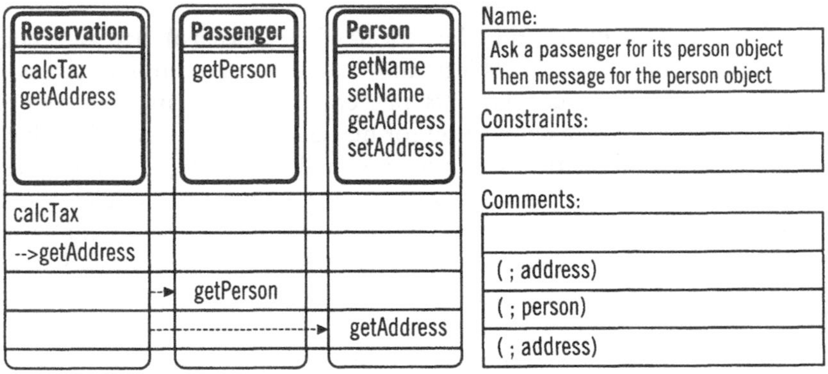

Figure 6.12   Asking for what you need (proxy scenario).

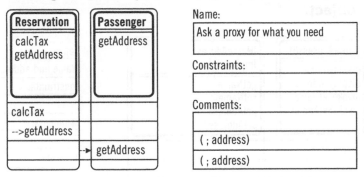

Figure 6.13   The object model for a proxy scenario.

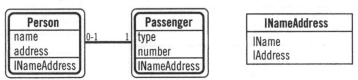

Person and Passenger with a single, combined interface

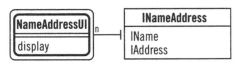

Each name–address user interface object is composed of a collection of INameAdress objects.

Finally, consider a NameAddressUI object. It's a user interface (UI) object, one that contains a number of smaller, hand-crafted or GUI builder-generated user interface objects such as text fields, buttons, scrollable lists, and the like. More important (from an object-modeling perspective), a NameAddressUI object knows some number of objects in classes that implement the INameAddress interface. NameAddressUI is not hardwired to objects in just one class. Instead, it works with objects from any class that implements the INameAddress interface.

Interfaces change the very nature of object connections — and get all of us to rethink what object modeling is all about. In Figure 6.13, a user interface object holds a collection of objects from any class that implements a specific interface. This shifts an object model builder's attention to what behavior that object must provide rather than what limit of class(es) of objects he or she should implement.

With interfaces, an object model gains better abstraction and simpler results. Take a look at a corresponding scenario view in Figure 6.14. It shows that interfaces change the very nature of a scenario, a time-ordered sequence of object interactions, and get all of us to rethink what scenarios are all about.

Figure 6.14   Name-address UI object interacting with `INameAddress`
object.

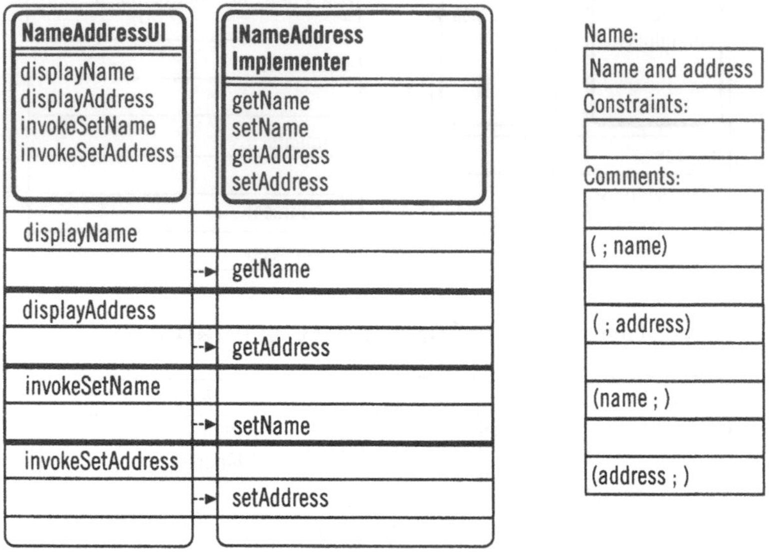

In the scenario shown in Figure 6.14, a user interface object sends a message to any object in a class that implements the needed interface. It doesn't matter where the receiving object's class is in the class hierarchy, or if its class spells out a different implementation (time vs. size tradeoffs will always be with us).

With interfaces, your attention shifts from what class of objects you're working with to what the interface is and what behavior you need from the object you're working with (now or in the future). Thus, you spend more time thinking about the behavior you need, rather than who might implement it.

Each scenario view in an interface delivers more impact within each scenario and reduces redundancy across related scenarios. The impact is: reuse within the current application and a greater likelihood of reuse in future applications.

Factor Out for Future Expansion

You can use interfaces as a futurist, too. What if you are wildly successful on your current project? Simply put: the reward for work well done is more work. So what is next? What other objects might you deal with in the future? Would they be objects that can plug in more easily, if you could establish a suitable interface right away?

You can add interfaces to improve model understanding now — and point to flexibility for future changes. You can also demonstrate to your customer that your model is ready for expansion. Perhaps the reward for doing so will actually be more money!

Figure 6.15 illustrates a sensor monitoring system, consisting of zones (areas within a building, for example) and sensors (temperature and motion measurement devices, for example). You can adjust the object model so a zone holds a collection of `IActivate` objects, as shown. To add to and remove from that collection, you add method signatures to the interface. Figure 6.16 shows a corresponding scenario view, showing add, activate, and deactivate.

Figure 6.15   Object model of a sensor monitoring system.

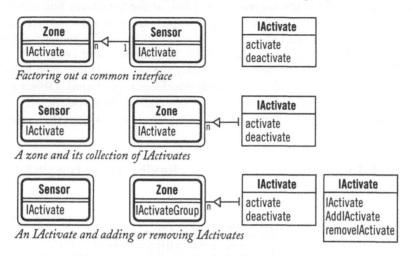

Figure 6.16   An `IActive` **interacting with its** `IActives`.

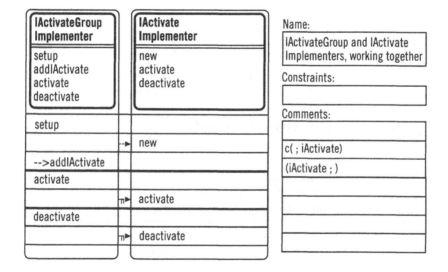

Flexibility, Extensibility, and Pluggability

One aspect of flexibility, extensibility, and pluggability is being able to combine objects that you are already working with in new ways — combinations you might not have anticipated. A zone could be a collection of other zones, and those zones could be a collection of sensors. A sensor could be a collection of other sensors. Nice.

Note that you could take a sensor object and give it a collection of zones, yet that makes little sense. Interfaces let you express the kind of behavior that must be supported. You must be reasonable when you decide what to plug together, though!

Another aspect of extensibility is being able to add new classes of objects: ones you can anticipate; and ones that may surprise you. Look at the interfaces you are establishing and consider what other classes of objects might implement that same interface. For zones and sensors, you might look ahead to additional IActivates, such as switches, motors, conveyor belts, and robot arms, as shown in Figure 6.17. Just add the new IActivate implementers to the object model. The scenario view stays exactly the same as before — no change required.

When all is said and done, designing with interfaces is profound. Object models are pluggable, no longer required to be hardwired by associations to a specific class. Scenario views are pluggable, too, no longer limited to the specific classes within that scenario. Have fun with interfaces!

Figure 6.17   Adding new IActivates — flexibility, extensibility, pluggability.

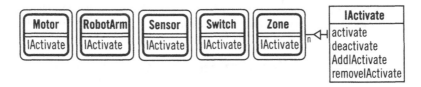

This article was adapted from the book, *Java Design: Building Better Apps and Applets* (Prentice Hall, 1997). This adaptation is included with the permission of Prentice Hall.

The authors thank Jon Kern of Lightship Inc. for his review and helpful feedback.

6.7.6 "An Introduction to Patterns"
by Scott W. Ambler, July 1998

Understanding the four types of patterns — design, analysis, process, and architectural — will help you solve some of the most common development problems.

One of the most important software concepts to be introduced during the 1990s is patterns. Patterns are an integral part of modern object-oriented development and a solution to a common problem. They provide a high level of reuse to developers: the reuse of the experience and thinking of another developer. Although a lot of good work has been done to identify and document patterns, little has been done to determine the best way to introduce developers to patterns — until now.

The purpose of this article is to guide your initial learning efforts. First, I would like to differentiate between four types of patterns: design patterns that document solutions to design problems; analysis patterns that describe solutions to business/domain problems; process patterns that describe proven, successful approaches for developing software; and architectural patterns that describe solutions to high-level design problems. Second, it is important to understand that you need to have a fairly good grasp of object technology before you can be effective with patterns. I typically suggest six to twelve months experience. Once you're ready, here is where you can start.

Ten Patterns to Get You Started

The following is an alphabetized, beginner's list of patterns. These patterns are easy to learn, yet still provide significant benefits. They are taken from several sources, and will introduce you to the wealth of knowledge encapsulated within patterns. Each item in the list is followed by the source from which it came, in parentheses. A complete list of the sources is listed at the end of this article.

Contact Point (Ambler)

I've beaten this analysis pattern to death in my columns during the past year or so. It basically shows that concepts such as surface addresses, phone numbers, and e-mail logons are similar — they are points of contact that your organization has with the business entities it interacts with. The strength of this pattern is that it improves the flexibility of the systems you build. For example, it simplifies giving your customers the choice of how they want to get their invoices: via the mail, fax, or e-mail.

Facade (Gamma et al.)

This design pattern provides a unified interface to a set of classes in a subsystem. It lets you reduce the coupling between subsystems and increases the extensibility of your software. For example, although a customer subsystem may be comprised of 50 classes, a single facade class called `CustomerCare` could be developed that would provide a single interface to other non-customer classes. Next month, I will show how you can use this design pattern to support both Microsoft's DCOM and CORBA in the same system.

Item-Line Item (Coad)

Item-Line Item is a design pattern that describes a strategy for normalizing the behavior required to describe an object. I applied it in "Implementing an Object-Oriented Order Screen" (Thinking Objectively, June 1998) when I described how to build an `Order` class: an `Order` object contains a collection of `OrderItem` objects that are described by `ItemDescription` objects. `OrderItem` objects know how many things were ordered and `ItemDescription` objects know the name, part number, and retail price of individual items that can be ordered. The advantage of this pattern is that it provides a common approach for a single object, in this case instances of `ItemDescription`, to describe key information about many specific objects, in this case instances of `OrderItem`.

Layers (Buschmann et al.)

This architectural pattern describes a proven approach to structuring software. An example of it is the Open System Interconnection Seven-Layer network architecture (the TCP/IP protocol used by the Internet conforms to the bottom four layers of this architecture). A second example is the class-type architecture approach, which states you should organize your classes into four layers: user interface, business/domain, persistence, and system. This pattern promotes development of software that is easy to extend and maintain.

Observations and Measurements (Fowler)

An analysis pattern, Observations and Measurements describes a generic way to manage a wide range of information about objects. Its basis is that the data attributes of a business object can be implemented as a measured observation. For example, the gender of a person is either male or female and that person's height can be within a given range of values (the pattern also supports the definition of unit objects such as inches or centimeters). This pattern helps you define observations for classes via the use of metadata, dramatically increasing the flexibility of your applications.

Organization Supertype (Fowler)

This analysis pattern describes a way to model the organizational structure of your company. It can be a key enabler of internal reporting systems and your security and access framework.

Proxy (Gamma et al.)

Proxy is a design pattern that describes how to build a surrogate or placeholder for another object that controls access to that object. Proxy objects are commonly used to provide access to an object in another memory space. For example, a client application may have a proxy object for an object that exists on a server, enabling the client application to collaborate with the object as if it existed on the client.

Reuse First (Ambler)

This process pattern states that when you need something you should first try to reuse an existing item that you currently own. If that fails, you should attempt to purchase a publicly available solution. Only as a last resort would you build the item you need. Although incredibly simple, the strength of this pattern is that it provides your organization with a common, reusable process for obtaining reuse at all levels of development. This includes project management, modeling, programming, and testing. It's a reusable process pattern that supports reuse — you can't help but like it.

Singleton (Gamma et al.)

This design pattern describes a strategy for ensuring that a class has a single instance that can be easily accessed within your software. A perfect example of a Singleton is a user profile object that indicates who is currently logged on to a machine, what their security access rights are, and what their preferences are.

State (Gamma et al.)

This design pattern lets an object change its behavior when its internal state changes, effectively making it appear that the object has changed its class. The strength of this pattern is that it supports polymorphism via collaboration, providing a mechanism for objects to offer different sets of behavior based on their state.

For example, in the domain of a car dealership, a vehicle exhibits different behavior depending on whether it is new, has been leased, has been sold, or is being used as a demo model. Instead of developing a Vehicle hierarchy (shown as a UML class diagram in Figure 6.18) consisting of the classes Vehicle (abstract) and various subclasses, you would develop a single concrete class called Vehicle that would know its associated VehicleState objects, as shown in Figure 6.19. With the first strategy, when a new vehicle is sold, you would need to change it into an instance of SoldVehicle — potentially a lot of work considering you need to update the associations with the other objects with which it's involved.

With the state pattern approach, you would merely need to associate the Vehicle object with a different VehicleState object — you don't have to copy the object or update any associations it is involved in (other than with the new VehicleState object). Another advantage of the state pattern is that it is easy to develop a history of an object. InFigure 6.19, a Vehicle has many VehicleState objects, only one of which is current. The rest represent the history of the state of the Vehicle object.

Figure 6.18   Implementing a car dealership.

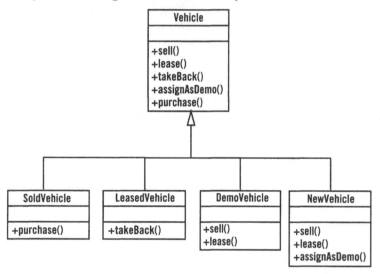

Figure 6.19   Applying the state design pattern to the car dealership.

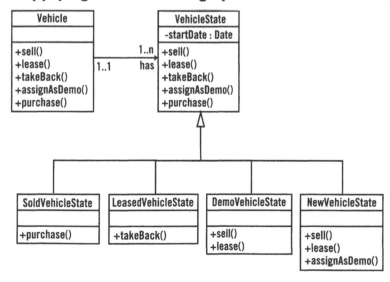

Getting the Most from Patterns

Patterns provide a high-level source of reuse that you can apply to your software development efforts. Although it is tempting to try to identify new patterns immediately, my advice is that you should walk before you run — gain several years of experience reusing existing patterns before you start identifying them yourself. Alternatively, it's often difficult to identify the forest pattern from all of the tree objects, so leave it up to the experienced foresters, lest you get lost in the pattern wildernesss.

To aid your pattern user efforts, look at Blueprint Technologies' Pattern Applicator add-on for Rational Rose, a tool that provides support for automatically applying patterns into your object-oriented models (www.blueprint_technologies.com). The hardest thing to understand about patterns is that not everything is a pattern — if this were true, we'd all be out of work by now. The reality is that patterns are one more technique to add to your development toolkit. It is a technique that provides an opportunity for spectacular levels of reuse within your software.

Key Patterns Resources — Recommended Reading

- *Analysis Patterns: Reusable Object Models* by Martin Fowler (Addison-Wesley, 1997)
- *Design Patterns: Elements of Reusable Object-Oriented Software* by Erich Gamma, Richard Helm, Ralph Johnson, and John Vlissides (Addison-Wesley, 1995)
- *Object Models: Strategies, Patterns, and Applications* by Peter Coad, David North, and Mark Mayfield (Prentice Hall, 1995)
- *Pattern-Oriented Software Architecture — A System of Patterns* by Frank Buschmann, Regine Meunier, Hans Rohnert, Peter Sommerlad, and Michael Stal (John Wiley and Sons, 1996)
- *Process Patterns: Delivering Large-Scale Systems Using Object Technology* by Scott Ambler (SIGS Books/Cambridge University Press, 1998)
- The Patterns web site: http://hillside.net/patterns/patterns.html

---

Chapter 7

Best Practices for the Test Workflow

The purpose of the Test workflow is to verify and validate the quality and correctness of your system. During the Elaboration phase, your goals are to stress test the technical prototype, to inspect and/or review your key artifacts (requirements model, design model, software architecture document), and to provide your test results as input into your project viability assessment at the end of the phase.

There are several important best practices that you should apply with respect to the test workflow. First, recognize that if you can build something, you can test it; you can inspect requirements, review your models, test your source code. In fact, I would go so far as to claim that if something isn't worth testing, then it likely isn't worth developing. Second, recognize that silence isn't golden, that your goal is to identify potential defects, not to cover them up. To be successful at testing you need to have the right attitude: that it is good to test and that it is even better to find defects.

If it isn't worth testing, then it isn't worth creating.

Third, you want to test often and test early. There are two reasons for this: a) we make most of our mistakes early in the life of a project, and b) the cost of fixing defects increases exponentially the later they are found. Technical people are very good at technical things such as design and coding — that is why they are technical people. Unfortunately technical people are often not as good at non-technical tasks such as gathering requirements and performing analysis — probably another reason why they are technical people. The end result, as shown in Figure 7.1, is that developers have a tendency to make more errors during the Requirements and Analysis and Design workflows than the Implementation, Test, and Deployment workflows.

Most defects are introduced early in the development of an application.

Figure 7.1   The decreasing probability of introducing defects.

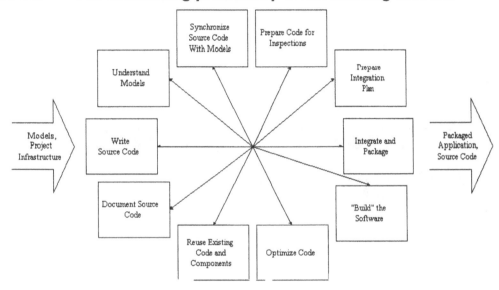

Figure 7.2 shows that the cost of fixing these defects rises the later they are found. This happens because of the nature of software development — work is performed based on work performed previously. For example, the work of the Analysis and Design workflow is performed based on the information gathered during the Requirements workflow. The Implementation workflow is done based on the Design Model developed as part of the Analysis and Design workflow. If a requirement was misunderstood, all modeling decisions based on that requirement are potentially invalid, all code written based on the models is also in question, and the testing efforts are now verifying the application against the wrong conditions. If errors are detected late in an iteration, or even in a future iteration, they are likely to be very expensive to fix. If they are detected during requirements definition, where they are most likely to be made, then they are likely to be much less expensive to fix (you only have to update your Requirements Model).

The cost of fixing defects rises exponentially the later they are found during the development of an application.

Figure 7.2   The rising costs of finding and fixing defects.

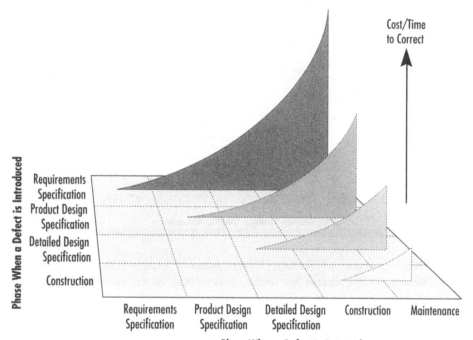

In Figure 7.3, you see a depiction of the *Full Lifecycle Object-Oriented Testing* (FLOOT) process pattern (Ambler, 1998a; Ambler, 1998b; Ambler, 1999) for testing and quality assurance indicating techniques that can be applied to both the Implementation and Test workflows. Although testing and quality assurance go hand-in-hand, they are two distinct tasks. The purpose of testing is to determine whether or not you built the right thing, and the purpose of quality assurance (QA) is to determine if you built it the right way. In section 7.1.1 "Object Testing Patterns" (*Software Development*, July 1999) I present an overview of the *Test In The Small* (Ambler, 1998b) and the *Test In The Large* (Ambler, 1999) process patterns that encapsulate the techniques of the FLOOT techniques. Process patterns are similar conceptually to design patterns; design patterns present proven solutions to common design problems whereas process patterns present proven solutions to common process problems. You use object-oriented and component-based development techniques because the software that you are developing is significantly more complex than the software you developed following the structured/procedural paradigm. Therefore, doesn't it make sense that testing that software would also be more complex? Of course it does. In this article I present a collection of testing best practices that can be directly applied to the Test workflow.

You apply the object and component paradigms because the software that you are developing is more complex. Complex software requires a more sophisticated approach to testing.

Figure 7.3   Full Lifecycle Object-Oriented Testing (FLOOT).

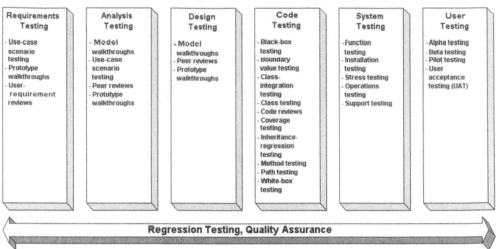

Robert Binder, author of *Testing Object-Oriented Systems: Models, Patterns, and Tools* (Addison-Wesley, 1999), shares his testing experiences in the section 7.1.2 "Scenario-Based Testing for Client/Server Systems" (*Software Development*, August 1995). Binder argues that user scenarios, effectively instances of use cases, should drive your user testing efforts. Requirements-driven testing is a fundamental best practice of the Unified Process, and Binder provides alternative but complementary advice for doing exactly that. To increase the effectiveness of your testing efforts, he suggests a technique for ordering your test cases — an important tactic considering you are rarely given the resources to completely test your systems. He also argues for the importance of maintaining traceability throughout your development efforts, a fundamental goal of the Configuration and Change Management workflow described in detail in volume 3, *The Unified Process Construction Phase* (Ambler, 2000a). Although Binder focuses on the testing of client/server (C/S) systems, the article presents fundamental testing concepts that are applicable regardless of the underlying development technology.

Requirements-driven acceptance and function testing is a fundamental best practice regardless of the underlying development technology.

In section 7.1.3 "Software Quality at Top Speed" (*Software Development*, August 1996) Steve McConnell, author of *Rapid Development: Taming Wild Software Schedules* (Microsoft Press, 1996), presents a collection of development best practices that focus on testing and quality assurance. He points out that projects that achieve the lowest defect rates also achieve the shortest schedules, implying that investment in testing pays off in the long run and that cutting corners during development will come back and haunt you. There are no shortcuts in software development. McConnell shares several key insights into effective testing including the need to test often and test early, the fact that inspections and reviews work very well, and how to do it right and do it once.

Projects that achieve the lowest defect rates also achieve the shortest schedules. Quality pays.

7.1 The Articles

7.1.1 "Object Testing Patterns" by Scott W. Ambler, July 1999
7.1.2 "Scenario-Based Testing for Client/Server Systems" by Robert Binder, August 1995
7.1.3 "Software Quality at Top Speed" by Steve McConnell, August 1996

7.1.1 "Object Testing Patterns"

by Scott W. Ambler, July 1999

Process patterns can help you test your object-oriented applications more thoroughly.

One of my fundamental philosophies is that if you create software that isn't important enough to test and validate, then you shouldn't have invested the time developing it in the first place. Testing is a critical part of the software process, just as critical as requirements gathering, analysis and design, implementation, deployment, and project management.

Like these other software development aspects, testing is a complex and time-consuming task that requires great skill and knowledge to be successful. Yet when you go to any bookstore's computer section, you'll see numerous programming, analysis and design, and project management books. But testing books? Nary a one. You'll find similar results at your local magazine rack, and finding a testing training course is difficult at best. In last month's column, I mentioned that the Rational Unified Process sorely lacks a comprehensive process for testing. Why is this the case? My theory is that although testing is important, it isn't sexy. In this month's column, I will discuss the often-neglected subject of object-oriented testing and how to integrate it with the Rational Unified Process (RUP).

Figure 7.4 depicts the RUP life cycle, an instance of the Unified Process. The figure shows the serial project phases across the top of the diagram and the workflows across the left side of the diagram. The good news is that testing is one of the RUP's core workflows. The bad news is that when you drill down to the workflow's activities, you find that it isn't complete. One way you can adapt the RUP to your organization's real-world needs is by applying process patterns — proven solutions that describe software development approaches in a given context. I'll provide an overview of several process patterns that describe testing techniques for object-oriented systems.

Figure 7.4   The Rational Unified Process life cycle.

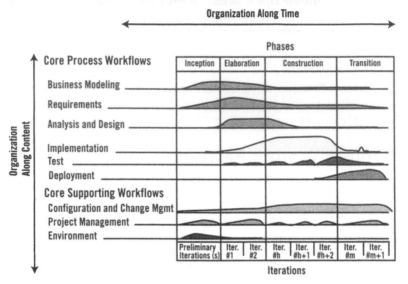

Figure 7.5 depicts the solution to the Test in the Small process pattern, which describes how to validate your work during the development process. You perform this pattern in parallel to your requirements, modeling, and implementation efforts as its techniques (described in the sidebar) validate the deliverables of these tasks. This pattern includes traditional quality assurance techniques such as model walkthroughs and user interface prototype reviews, traditional unit testing approaches such as method and function testing, and newer approaches such as class integration and inheritance regression testing.

I believe these Test in the Small techniques should be applied to the RUP's Inception, Elaboration, and Construction phases. As shown in Figure 7.4, the RUP does not contain testing during Inception, which I find surprising. Think about it, one goal of this phase is to identify the initial requirements for your system, requirements that you can easily validate using several techniques from this pattern. You create several artifacts during the Inception phase, including an initial use case model and likely an initial user interface model that you can and should test. Therefore, developers need a couple of "bumps" on the Test workflow during the Inception phase.

The Test in the Large process pattern, shown in Figure 7.6, depicts testing techniques (described in the sidebar on page 240) that you are likely to employ during the Transition phase. These are familiar techniques you have employed for years, such as stress testing and operations testing, to validate your software just before deploying it to your users. This process pattern has two parts: the system test portion in which your test engineers put your software through its paces, and user test in which your user community tests your software.

Figure 7.7 depicts the Technical Review process pattern, showing the potential tasks you need to perform for a successful model walkthrough, code inspection, document walkthrough, or prototype review. Each of these tasks holds the same basic process, although a review is generally more formal than an inspection, which in turn, is more formal than a walkthrough. Although Technical Review is arguably a quality assurance process pattern, I've included it in this article because the RUP does not include a separate quality assurance work-flow.

Figure 7.5   The solution to the Test in the Small process pattern.

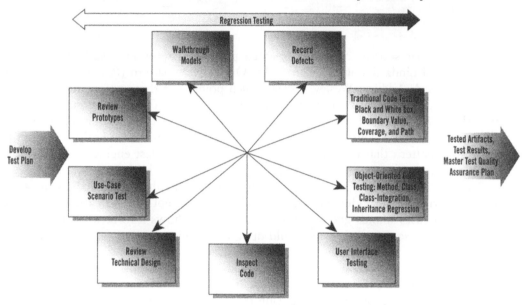

Figure 7.6   The solution to the Test in the Large process pattern.

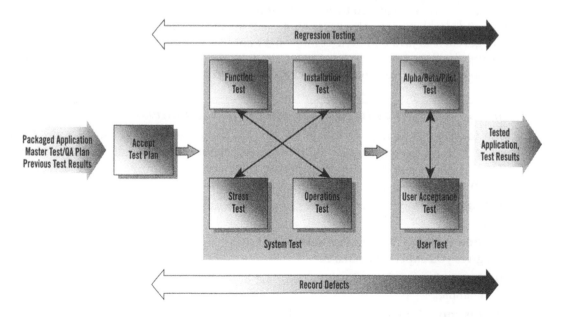

Figure 7.7   The Technical Review process pattern.

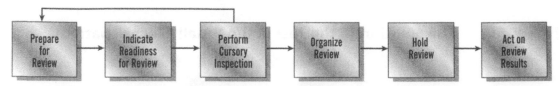

Several other sources of testing-oriented process patterns are available, including D.E. DeLano's and Linda Rising's Designers Are Our Friends pattern (*Pattern Languages of Program Design 3*, Addison-Wesley, 1998), which points out that test engineers need to work with developers to identify and solve software problems. This pattern supports the fundamental concept that quality is everyone's responsibility and that your entire team must work effectively together if your software is to be successful. Get Involved Early is another DeLano and Rising pattern that is dear to my heart. It argues that test engineers should be actively involved in the software development process right from the beginning.

Not only can you use process patterns to extend and enhance the RUP, you can also use them to clarify the misunderstandings in it. The RUP concept that use cases drive the development of test cases sounds good in theory, but practice shows it to be inaccurate. At best it is partially true, use cases can and should drive the development of function and user acceptance test (UAT) cases, but there is significantly more to object-oriented testing than just these two testing forms. At worst, this concept has confused our industry by supporting the argument that "you can't test objects," because it's obvious that you need to do far more than function and user acceptance testing. Unfortunately, this confusion has driven many organizations away from object technology completely.

Object-oriented testing is a complex endeavor, far more complex than structured testing. Further, it is virtually ignored by most of our industry in favor of sexier topics such as component-based development, object-oriented analysis and design, and cool new programming languages such as Java and Perl. Testing might not be glamorous, but it's an important part of software development that your organization simply can't ignore. To paraphrase a well-known motorcycle gang: "Test to Live; Live to Test."

Common Approaches to Testing Object-Oriented Software

Alpha testing   A testing period in which pre-release versions of software products are released to users who need access to the product before it is officially deployed. In return, these users will report any defects to the software developers. Alpha testing is typically followed by beta testing.

Beta testing   A similar process to alpha testing, except the software product should be less buggy.

Black-box testing   Testing that verifies that given input A the component or system being tested gives you expected results B.

Boundary-value testing   Testing of unusual or extreme situations that your code should be able to handle.

Class testing   Testing that ensures a class and its instances (objects) perform as defined.

Class-integration testing   Testing that ensures an application's classes, and their instances, perform as defined.

Code inspection   A form of technical review in which the deliverable being reviewed is source code.

Coverage testing   Testing that ensures all lines of code are exercised at least once.

Function testing   A part of systems testing in which development staff confirm that their application meets the specified user requirements.

Inheritance-regression testing   Running the test cases of all the super classes, both direct and indirect, on a given subclass.

Installation testing   Testing that ensures your application can be installed successfully.

Integration testing   Testing that verifies that several portions of software work together.

Method testing   Testing that verifies a method (member function) performs as defined.

Model review   A technical review in which a model is inspected.

Operations testing   Testing that ensures the needs of operations personnel who have to support or operate the application are met.

Path testing   Testing that ensures all logic paths within your code are exercised at least once.

Peer review   A style of technical review in which a project artifact, or portion thereof, is inspected by a small group of experts.

Pilot testing   A testing process equivalent to beta testing that organizations use to test applications they have developed for internal use.

Proof-of-concept prototyping   Writing software to prove or test a technology, language, environment, or approach's viability. Also called technical prototyping.

Prototype walkthrough   The process in which your users work through a collection of use cases using a prototype as if it were the real thing. Its main goal is to test whether or not the prototype's design meets their needs.

Regression testing   Testing that ensures previously tested behaviors still work as expected after changes have been made to an application.

Stress testing   Testing that ensures the system performs as expected under a high volume of transactions, high number of users, and so on. Also referred to as load testing.

Support testing   Testing that ensures the needs of support personnel who have to support the application are met.

System testing   A testing process in which you find and fix any known problems to prepare your application for user testing.

Technical review   A testing technique in which the design of your application is examined critically by a group of your peers. A review will typically focus on accuracy, quality, usability, and completeness. This process is often referred to as a walkthrough, inspection, or a peer review.

Use-case scenario testing   A testing process in which users work through use cases with the aid of a facilitator to verify that a user interface prototype fulfills the needs of its users and that the identified classes for a system fulfill the requirements described in the use cases.

User testing   Testing processes in which the user community, as opposed to developers, performs the tests.

User-interface testing   The testing of the user interface to ensure that it follows accepted standards and meets its requirements. User-interface testing is often referred to as graphical user interface (GUI) testing.

Volume testing   A subset of stress-testing that deals specifically with determining how many transactions or database accesses an application can handle during a defined period of time.

White-box testing   Testing that verifies specific lines of code work as defined.   Also referred to as clear-box testing.

7.1.2 "Scenario-Based Testing for Client/Server Systems"

by Robert Binder, August 1995

Scenario-based testing tells you where to focus your quality assurance efforts and when it's time to stop.

Testing is a game of trade-offs, not unlike clearing a mine field. You can't be sure where all the mines are buried, but you want to find and detonate as many as possible before you send the troops across the field. Since you can't dig up every square yard, you need a plan that gives you the best chance of finding the mines.

You don't want to waste time on ground you've already cleared. You'd like to have a strategy for searching that lets you know when to stop looking. In software, bugs (faults) are like mines. A testing strategy is like a map of the mine field with a search path. Coverage is like marking off the paths you've checked.

Testing Paths

Many code-based coverages have been studied and used; statement, decision, and dataflow coverage are well-known. Statement coverage shows the percent of source code statements executed during a test. If all statements have been executed at least once, you get 100% statement coverage.

But statement coverage is by no means clearing the mine field. Many faults can go undetected and are waiting for an unsuspecting user. Regardless of the testing strategy, the motivation for coverage is the same: you want to get an objective measurement of how much has been tested so you can manage the test process.

Today's client-server applications can be large and complex. It is not unusual to find several technologies in a single sys-tem: GUI generators, GUI languages, third-generation languages, fourth-generation languages, object-oriented languages, and database command languages. With this new mix, conventional code coverage is not as useful in telling us when to stop looking for mines. More than ever, you need some systematic way to decide when you've done an adequate job of testing. The need is greater because the technology is inherently more complex and therefore more fault prone.

The operational profile derived from user scenarios provides a solution. In the article "Operational Profiles in Software-Reliability Engineering" (*IEEE Software*, Mar. 1993), John D. Musa describes his scenario-based testing approach, which I use in my Client/Server Systems Engineering methodology.

The Operational Profile

An operational profile offers a frame-work for developing test cases and man-aging testing. This approach can pro-vide a technology-independent coverage measure. An operational profile has two main parts: usage scenarios and scenario probability.

The notion of the "average" or "typical" user can be misleading. Different people use a system differently. But information about different use patterns (scenarios) is often overlooked in formal and informal testing. Formal test plans are derived from code, requirements, or checklists. Informal testing is usually testing by poking around and often becomes a demonstration of features a developer is particularly proud of. User-driven testing (acceptance, beta, and so on) also involves a lot of poking around. User tests are often selected out of curiosity or fear. For example, a new lookup facility providing live action video may get a lot of use (curiosity). If a "remote update" caused data corruption in the past, the user will probably try to repeat the problem (fear).

Explicit definition of scenarios avoids the haphazard results of testing by poking around. A scenario identifies usage categories — the people who use the system and the different ways in which they use it. To create an accurate scenario, you need to know about users, the activities they perform, and the environments in which they perform these activities. In other words, you need to define the who, what, where, how, and why behind a system.

I'll illustrate an operational profile with a hypothetical example: the Acme Widget Company Order System (Acme Widget supplies Wiley Coyote with Road-Runner-catching equipment). To begin, say 1,000 people use the Acme Widget order system. Their use patterns differ according to how often they use the system. Of the group, 300 are experienced, about 500 use the system on a monthly or quarterly basis, and the balance use the system less than once every six months. This profile is shown in Table 7.1.

Table 7.1   Operational profile for Acme Widget Order System.

User Type	% of Total	Location	% of Total	Activity	% of Total	Scenario Probability
Experienced	0.3	Plant	0.80	Inquiry	0.05	0.0120
	0.3	Plant	0.80	Update	0.05	0.0120
	0.3	Plant	0.80	Print Ticket	0.90	0.2160
	0.3	Office	0.01	Order Entry	0.70	0.0210
	0.3	Office	0.01	Update	0.20	0.0060
	0.3	Office	0.01	Inquiry	0.10	0.0030
	0.3	Customer Site	0.01	Order Entry	0.10	0.0030
	0.3	Customer Site	0.01	Update	0.20	0.0060
	0.3	Customer Site	0.01	Inquiry	0.70	0.0210

User Type	% of Total	Location	% of Total	Activity	% of Total	Scenario Probability
Cyclical	0.5	Plant	0.10	Inquiry	0.05	0.0025
	0.5	Plant	0.10	Update	0.05	0.0025
	0.5	Plant	0.10	Print Ticket	0.90	0.0450
	0.5	Office	0.50	Order Entry	0.30	0.0750
	0.5	Office	0.50	Update	0.20	0.0500
	0.5	Office	0.50	Inquiry	0.50	0.1250
	0.5	Hand Held	0.40	Order Entry	0.95	0.1900
	0.5	Hand Held	0.40	Update	0.02	0.0040
	0.5	Hand Held	0.40	Inquiry	0.03	0.0060
Infrequent	0.2	Plant	0.05	Report	0.75	0.0075
	0.2	Plant	0.05	Update	0.15	0.0015
	0.2	Plant	0.05	Inquiry	0.10	0.0010
	0.2	Office	0.95	Inquiry	0.60	0.1140
	0.2	Office	0.95	Update	0.10	0.0190
	0.2	Office	0.95	Report	0.30	0.0570
						1.0000

The use environment of a system may span a wide range of conditions. This is the "where" part of a scenario. You want to consider anything that would influence use or the system's ability to perform. For client/server applications, this could be sites, operating units, or departments where the system will be used. For systems designed to run on various platforms or network configurations, each physically distinct target constitutes an environment. Time can also define different environments — day vs. night, winter vs. sum-mer, and so on.

Suppose Acme Widget has several locations with significantly different use patterns: plant, office, customer site, and hand-held access (devices used by delivery, sales, and service people on the road). Some locations are only visited by certain users. For example, only experienced users go to customer sites.

Next, you consider activities — the "what." An activity is defined from the user's point of view. It is a distinct, definable interaction with the system. It is part of a task or procedure the user must do. An activity is not the same as a system function or feature. It is an identifiable pattern of system use. It must have specific time characteristics (the "when"): interval, cycle, duration, frequency, and so on.

Activities typically vary by user and environment. Suppose the main activities for Acme Widget are order entry, order inquiry, order update, printing a shipping ticket, and producing periodic reports. After studying the use patterns, you find different proportions of user type and location.

For example, the infrequent user will never print a shipping ticket but is likely to request periodic reports. The answers can be arranged many ways to determine scenarios, but they should be organized into a hierarchy.

How Probable Is It?

Next, calculate the probability of each scenario. Simply estimate the relative frequency or probability of each factor. You do not need extreme precision in these estimates. However, the estimates must be proportionally correct, from most to least frequent.

Multiply the factor proportions to get the scenario's probability. For example, looking at the first factor in Table 7.1, about 30% are experienced users. The second factor (location) shows that 80% of the experienced group uses the system in the plant. Nine times in ten, an experienced user in the plant will use the system to print a shipping ticket. The column on the far right gives the probability of this scenario: 0.216. So, of all uses of the system, about 22% will be by experienced users in the plant to print a shipping ticket. The individual scenario probabilities must add up to 1.0.

Using the Profile

Now that you have the operational pro-file for the Acme Widget Order System, you can use it to assign testing priority and measure coverage. The sorted scenarios are a key part of the operational profile strategy. You select scenarios to test in order of probability; most frequent first, least frequent last, as shown in Table 7.2.

Table 7.2   Prioritized operational profile.

User Type	% of Total	Location	% of Total	Activity	% of Total	Scenario Probability
Experienced	0.3	Plant	0.80	Print Ticket	0.90	0.2160
Cyclical	0.5	Hand Held	0.40	Order Entry	0.95	0.1900
Cyclical	0.5	Office	0.50	Inquiry	0.50	0.1250
Infrequent	0.2	Office	0.95	Inquiry	0.60	0.1140
Cyclical	0.5	Office	0.50	Order Entry	0.30	0.0750
Infrequent	0.2	Office	0.95	Report	0.30	0.0570
Cyclical	0.5	Office	0.50	Update	0.20	0.0500
Cyclical	0.5	Plant	0.10	Print Ticket	0.90	0.0450
Experienced	0.3	Office	0.10	Order Entry	0.70	0.0210
Experienced	0.3	Customer Site	0.10	Inquiry	0.70	0.0210
Infrequent	0.2	Office	0.95	Update	0.10	0.0190
Experienced	0.3	Plant	0.80	Update	0.05	1.0120
Experienced	0.3	Plant	0.80	Inquiry	0.05	0.0120
Infrequent	0.2	Plant	0.05	Report	0.75	0.0075
Cyclical	0.5	Hand Held	0.40	Inquiry	0.03	0.0060
Experienced	0.3	Customer Site	0.10	Update	0.20	0.0060
Experienced	0.3	Office	0.10	Update	0.20	0.0060
Cyclical	0.5	Hand Held	0.40	Update	0.02	0.0040
Experienced	0.3	Customer Site	0.10	Order Entry	0.10	0.0030
Experienced	0.3	Office	0.10	Inquiry	0.10	0.0030
Cyclical	0.5	Plant	0.10	Inquiry	0.05	0.0025
Cyclical	0.5	Plant	0.10	Update	0.05	0.0025
Infrequent	0.2	Plant	0.05	Update	0.15	0.0015
Infrequent	0.2	Plant	0.05	Inquiry	0.10	0.0010
						1.0000

This method maximizes operational reliability for a given testing budget and schedule. You test the most likely scenarios first because the most frequently used functions have the greatest impact on operational reliability. If a heavily used function is buggy, the system will fail frequently; if a rarely used function is buggy, the system will fail infrequently. Selecting tests (and removing bugs) according to usage frequency rapidly decreases failure frequency. If schedule and cost constraints mean you can't test all scenarios or must reduce the number of test cases, cutting or reducing low frequency scenarios will have the least negative effect on overall reliability.

The operational profile is a framework for a complete test plan. For each scenario, you need to determine which functions of the system under test will be used. Systems typically have thousands of individual functions (for example, all the object/action pairs in a GUI). A single scenario typically involves tens of system functions. You need to identify the specific system functions that are used in each scenario.

A scenario often involves several system functions; these are called runs. Each run is an event/response thread. It has an identifiable input and produces a distinct output. For example, the experienced/ plant/ticket scenario might be composed of several runs:

- Display pending shipments
- Display scheduled pickups
- Assign carrier to shipment
- Enter carrier lading information
- Print shipment labels
- Enter on-truck time stamp.

The next step is to prepare a detailed test case for each run. There are many ways to do this. At a minimum, try several typical input-output combinations and several exceptional combinations. If many runs and data conditions exist, you can identify high priority runs using the scenario-ranking technique. State-based testing may be useful for event-driven GUIs. A complete discussion of techniques used to prepare detailed functional test cases is beyond the scope of this article.

Some scenarios may be low probability but have high potential impact. For example, suppose Acme Widget is promoting order entry at the customer site as a key selling feature. Even though this accounts for only three in a thousand uses, it should be tested as if it were a high-priority scenario.

Integration Test Strategy

Operational profiles provide a way to design, plan, and manage system test, but they do not obviate the need for unit or integration testing. Scenario-based testing assumes a minimally usable system, so completion of integration test is a necessary precondition.

Integration testing for typical client/server applications may be challenging if you're targeting a multiplatform environment. A primary goal of client/server integration testing should be to reveal and correct interoperability problems.

Multiplatform integration requires a controllable test environment that spans all platforms in the target environment. However, debug and test tools are typically platform specific. Thus, it may be hard to perform controlled testing and debugging of system threads that span several platforms (for example, mainframe MVS and UNIX).

Careful planning for integration test support is important. Choose one platform to support your testing effort. Locate your primary test repository on this platform. Develop reliable inter-faces from your test repository to the test tools on each platform. If you haven't already set up and tested this kind of environment, plan extra time during integration testing to do so.

In addition to typical testing tasks, multiplatform integration testing often requires defining, loading, and synchronizing databases; securing access across gateways, bridges, and routers; and establishing name control. Coordination of production and test environments typically becomes more complex. A robust, multiplatform configuration management system is indispensable.

Integration testing of your application can proceed relatively smoothly with a carefully configured and stabilized system testing environment. At a minimum, your integration test suite should:

- Bring the entire system online
- Run a set of transactions that exercise all physical subsystem interfaces under normal conditions

- Run a set of transactions that exercise typical exceptions visible over physical subsystem interfaces (for example, server offline)
- Perform normal shutdown procedures
- Repeat this cycle at least once.

Your application and the test environment are now ready to support scenario testing. You can include stress and performance testing in the scenario framework, but this will require an analysis of the implementation architecture.

Traceability and Coverage

Test case development begins in parallel with requirements definition in the Client/Server Systems Engineering methodology. This process helps you validate requirements as you identify them. Developing scenarios (even with-out running them) is an effective strategy because it checks consistency among expectations, requirements, and the system under development.

To take advantage of this cross-checking, you need an explicit definition of system capabilities and a way to associate capabilities and scenarios. Establishing traceability simply means recording and analyzing these associations. Nearly all software development methodologies produce some kind of external definition of system capabilities, for example:

- Line item requirements
- Use cases
- GUI object-action pairs (menu choices, radio buttons, and so on)
- Feature and function points
- Business transactions
- Business rules implemented as stored procedures or triggers
- Essential activities
- Event and response list
- CRUD matrix entries
- Steps in user processes and procedures
- Features in product literature
- Features in system documentation

No matter which form of external definition you use, establishing trace-ability is a simple bookkeeping operation. For each run, note the external definition item exercised by that run. Clearly, every externally defined capability should appear in at least one run. If a capability does not have a corresponding run, you should develop a scenario or drop the capability. Each run should be associated with at least one externally defined capability. If no scenario for a capability exists, you need to determine why. Is the scenario inappropriate or has the scenario revealed a missing capability?

You should evaluate traceability as scenarios are developed and testing progresses. A spreadsheet or personal data-base can track this information. Support for traceability is also provided in some CASE and CAST tools. With trace-ability, you can quantify the extent to which scenarios have covered system capabilities as well as quickly identify inconsistencies.

Create Your Scenario

With the proper automated support, scenarios can be reused over a long period of time by many people. Effective scenario-based testing requires good planning and an appropriate investment in time, facilities, and human resources. For example:

- User and customer cooperation will probably be needed to identify realistic scenarios.
- Scenarios should be validated with users and customers.
- Scenario-based, test-case development and evaluation requires people with a high level of product expertise.
- These people are typically in short supply, but this is arguably a good use of their time.
- A large number of test cases can be produced. Therefore well-defined housekeeping procedures with auto-mated support are needed.

With multitechnology client-server systems, conventional coverage measures will be less useful as approximations of system reliability. The operational profile can provide an objective, use-based measurement of test coverage for such systems. With the operational profile and an estimate of testing cost in hand, managers and developers can make rational decisions about the best use of available resources.

Scenario-based testing will deliver more reliable systems compared to testing by poking around. It is a rational alternative to managing testing by the seat of your pants. Perhaps most important, it is driven by a real-world user perspective.

Scenarios, Use-Cases, and Quality Assurance

What is the number-one way to lower your costs and improve the quality of your delivered software? Imagining how your users will use it, writing that down, and keeping track of it. That's the short version of the whole scenario and use-case furor. Of course, there is a little more to it, and you'll want to consult the following resources from *Software Development* before delving into use-case and scenario testing.

What are use-cases and how are they different from scenarios?
- "The Mouse That Roared," Object Lessons, March 1995.

Designing from use cases
- "Use Case Scenario Testing," Scott Ambler, July 1995. (see also: volume 1, *The Unified Process Inception Phase*.)
- "Mining Use Cases For Objects," Object Lessons, May 1995.

Tracking requirements
- "Requirements Analysis and Prototyping: Fast, Cheap, and Robust," by Stephen Andriole, May 1994.
- "The A-Train to Product Management," Hello, World!, January 1994.

Testing
- "Software Testing: Concepts, Tools, and Techniques," by Steve Rabin, November 1993.
- "Testing Techniques for Quality Assurance," by Mark Gianturco, August 1994.
- "Buyers Guide: Quality Assurance Tools," March 1994.

7.1.3 "Software Quality at Top Speed"

by Steve McConnell, August 1996

Emphasizing quality measures might be the smartest thing you do to deliver your software faster, cheaper, and better.

Software products exhibit two general kinds of quality that affect software schedules in different ways. The first kind that people usually think of when they refer to "software quality" is low defect rate. The other kind includes all other characteristics that you think of when you consider a high-quality software product — usability, efficiency, robustness, maintainability, portability, and so on. In this article, I will focus on the relationship between the first kind of quality and development speed.

Some project managers try to shorten their schedules by reducing the time spent on quality assurance practices such as design and code reviews. Some shortchange the upstream activities of requirements analysis and design. Others try to make up time by compressing the testing schedule, which is vulnerable to reduction since it's the critical-path item at the end of the schedule.

These are some of the worst decisions a person who wants to maximize development speed can make. In software, higher quality (in the form of lower defect rates) and reduced development time go hand in hand. Figure 7.8 illustrates the relationship between defect rate and development time. As a rule, the projects that achieve the lowest defect rates also achieve the shortest schedules.

Figure 7.8   Relationship between defect rate and development time.

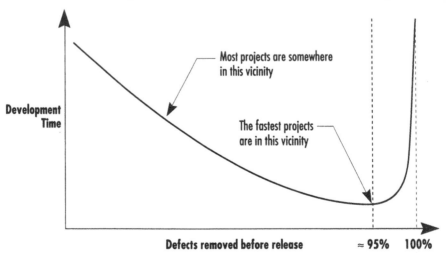

A few organizations have achieved extremely low defect rates, as shown on the far right of the curve in Figure 7.8. When you reach that point, further reducing the number of defects will tend to increase the amount of development time. This applies to life-critical systems such as the life-support systems on the space shuttle. It doesn't apply to the rest of us.

The rest of us would do well to learn from a discovery made by IBM in the 1970s: Products with the lowest defect counts also have the shortest schedules (Jones, 1991). Many organizations currently develop software with defect levels that give them longer schedules than necessary. After surveying about 4,000 software projects, Capers Jones reported that poor quality was one of the most common reasons for schedule overruns (Jones, 1994). He also reported that poor quality is implicated in close to half of all canceled projects. Additionally, a Software Engineering Institute survey found that more than 60% of the organizations assessed suffered from inadequate quality assurance (Kitson and Masters, 1993). On the curve in Figure 7.8, those organizations are to the left of the 95% removal line.

That 95% removal line — or some point in its neighborhood — is significant because that level of prerelease defect removal appears to be the point at which projects achieve the shortest schedules, least effort, and highest levels of user satisfaction (Jones, 1994). If you're finding more than 5% of your defects after your product has been released, you're vulnerable to the problems associated with low quality, and you're probably taking longer than necessary to develop your software.

Design Shortcuts

Projects that are rushed are particularly vulnerable to shortchanging quality assurance at the individual developer level. Any developer who has been pushed to ship a product quickly knows how much pressure there can be to cut corners because "we're only three weeks from shipping." For example, rather than writing a separate, completely clean printing module, you might piggyback printing onto the screen display module. You know that's a bad design, that it isn't extendible or maintainable, but you don't have time to do it right. You're being pressured to get the product done, so you feel compelled to take the shortcut.

Two months later, the product still hasn't shipped, and those cut corners come back to haunt you. You find users are unhappy with printing, and the only way to satisfy their requests is to significantly extend the printing functionality. Unfortunately, in the two months since you piggybacked printing onto the screen display module, the printing and screen display functionality have become thoroughly intertwined. Redesigning printing and separating it from the screen display is now a tough, time-consuming, error-prone operation.

The upshot is that a shortcut that was supposed to save time actually wasted it in the following ways:

- The original time spent designing and implementing the printing hack was completely wasted because most of that code will be thrown away. The time spent unit testing and debugging the printing hack code was also wasted.
- Additional time must be spent to strip the printing-specific code out of the display module.
- Additional testing and debugging time must be spent to ensure that the modified display code still works after the printing code has been stripped out.
- The new printing module, which should have been designed as an integral part of the system, has to be designed onto and around the existing system, which was not designed with it in mind.

All this happens when the only necessary cost — if the right decision has been made at the right time — is to design and implement one version of the printing module. And you still have to do that anyway.

This example is not uncommon. Up to four times the normal number of defects are reported for released products developed under "excessive schedule pressure" (Jones, 1994). Projects that are in schedule trouble often become obsessed with working harder rather than working smarter. Attention to quality is seen as a luxury. The result is that projects often work dumber, which gets them into even deeper schedule trouble.

Error-Prone Modules

Error-prone modules, which are modules responsible for a disproportionate number of defects, are a major factor in quality assurance — especially in rapid development. Barry Boehm reported that 20% of program modules are typically responsible for 80% of errors (Boehm, 1987). Similarly, on its IMS project, IBM found that 57% of errors clumped into 7% of the modules (Jones, 1991).

Modules with such high defect rates are more expensive and time-consuming to deliver than less error-prone modules. Normal modules cost about $500 to $1,000 per function point to develop (Jones, 1994). Error-prone modules cost about $2,000 to $4,000 per function point. Error-prone modules tend to be more complex, less structured, and much larger than other modules in the system. They often are developed under excessive schedule pressure and are not fully tested.

If development speed is important, make identification and redesign of error-prone modules a priority. Once a module's error rate hits about 10 defects per thousand lines of code, review it to determine whether it should be redesigned or reimplemented. If it's poorly structured, excessively complex, or excessively long, redesign the module and reimplement it from the ground up. You'll shorten the schedule and improve the quality of your product at the same time.

Quality Assurance and Development Speed

If you can prevent defects or detect and remove them early, you can realize a significant schedule benefit. Studies have found that reworking defective requirements, design, and code typically consumes 40% to 50% of the total cost of software development (Boehm, 1987; Jones, 1986). As a rule of thumb, every hour you spend on defect prevention will reduce your repair time from three to ten hours. In the worst case, reworking a software requirements problem once the software is in operation typically costs fifty to two hundred times what it would take to rework the problem in the requirements stage (Boehm and Papaccio, 1988). It's easy to understand why. A one-sentence requirement can expand into five pages of design diagrams, then into five hundred lines of code, fifteen pages of user documentation, and a few dozen test cases. It's cheaper to correct an error in that one-sentence requirement at requirements time than it is after design, code, user documentation, and test cases have been written to it.

Figure 7.9 illustrates the way that defects tend to become more expensive the longer they stay in a program.

The savings potential from early defect detection is huge — about 60% of all defects usually exist by design time, and you should try to eliminate them before then (Gilb, 1988). A decision early in a project not to focus on defect detection amounts to a decision to postpone defect detection and correction until later in the project when they will be much more expensive and time-consuming. That's not a rational decision when time is at a premium.

Figure 7.9   How long a defect goes undetected and the cost to fix it.

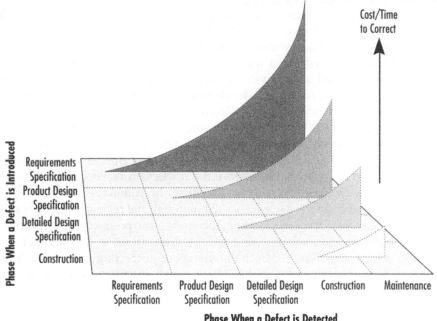

Testing

Various quality assurance measures have different effects on development speed. The most common quality assurance practice is undoubtedly execution testing, finding errors by executing a program and seeing what happens. The two basic kinds of execution testing are unit tests and system tests. In unit tests, the developer checks his or her own code to verify that it works correctly. In system tests, an independent tester checks to see whether the system operates as expected.

Testing is the black sheep of quality assurance practices as far as development speed is concerned. It can certainly be done so clumsily that it slows down the development schedule, but most often its effect on the schedule is only indirect. Testing can discover if the product's quality is too low for it to be released, and that the release has to be delayed until the product can be improved. Testing thus becomes the messenger that delivers bad news.

The best way to leverage testing from a rapid development viewpoint is to plan ahead for bad news — set up testing so that if there's bad news to deliver, it will happen as early as possible.

Technical Reviews

Technical reviews detect defects in requirements, design, code, test cases, and other work products. These reviews vary in level of formality and effectiveness, and they play a more critical role in maximizing development speed than testing does.

The least formal and most common kind of review is the walkthrough, which is any meeting at which two or more developers review technical work with the purpose of improving its quality. Walkthroughs are useful to rapid development because you can use them to detect defects earlier than you can with testing.

Code reading is a somewhat more formal review process than a walkthrough but nominally applies only to code. In code reading, the author of the code hands out source listings to two or more reviewers. The reviewers read the code and report any errors to the code's author. David Card reported that a study at NASA's Software Engineering Laboratory found that code reading detected about twice as many defects per hour of effort as testing (Card, 1987). That suggests that, on a rapid-development project, some combination of code reading and testing would be more schedule-effective than testing alone.

Inspections are the most formal kind of technical review, and they have been found to be extremely effective in detecting defects throughout a project. Developers are trained in the use of inspection techniques and play specific roles during the inspection process. The moderator hands out the material to be studied before the inspection meeting. The reviewers examine the material before the meeting and use checklists to stimulate their reviews. During the inspection meeting, the author paraphrases the material, the reviewers identify errors, and the scribe records the errors. After the meeting, the moderator produces an inspection report that describes each defect and indicates what will be done about it. Throughout the inspection process, you gather data about defects, hours spent correcting them, and hours spent on inspections. This is so you can analyze the effectiveness of your software development process and improve it.

Reviews find defects earlier, which saves time. They are more cost-effective on a per-defect-found basis because they simultaneously detect the symptom of the defect and its underlying cause.

Inspections have been found to produce net schedule savings of 10% to 30% because they can be used early in the development cycle (Gilb and Graham, 1993). One study of large programs found that each hour spent on inspections avoided an average of 33 hours of maintenance, and inspections were up to 20 times more efficient than testing (Russell, 1991).

Technical reviews are a useful and important supplement to testing. Reviews find defects earlier, which saves time. They are more cost-effective on a per-defect-found basis because they simultaneously detect the symptom of the defect and its underlying cause. Testing detects only the symptom of the defect; the developer still has to isolate the cause by debugging. Capers Jones reported that reviews also tend to find a higher percentage of defects (Jones, 1986). Reviews are also a time when developers share their knowledge of best practices with each other, which increases their rapid-development capability over time. Technical reviews are thus a critical component of any development effort that aims to achieve the shortest possible schedule.

When a software product has too many defects, developers spend more time fixing the software than they spend writing it in the first place. Most organizations have found that an important key to achieving the shortest possible schedules is focusing their development processes so that their work is done right the first time. "If you don't have time to do the job right," the old chestnut goes, "when will you find the time to do it over?"

Additional Reading The publications included in this list are also referenced throughout this article. Use them to your advantage when seeking further details about this subject.

- *Applied Software Measurement: Assuring Productivity and Quality* by Capers Jones (McGraw-Hill, 1991)
- *Assessment and Control of Software Risks* by Capers Jones (Yourdon Press, 1994)
- "An Analysis of SEI Software Process Assessment Results, 1987-1991" by David Kitson and Stephen Masters in *Proceedings of the Fifteenth International Conference of Software Engineering* (IEEE Computer Society Press, 1993)
- *Tutorial: Programming Productivity: Issues for the Eighties, 2nd Ed.* Capers Jones, ed. (IEEE Computer Society Press, 1986)
- "Improving Software Productivity" by Barry Boehm in *IEEE Computer*, Sept. 1987
- "Understanding and Controlling Software Costs" by Barry Boehm and Philip Papaccio in *IEEE Transactions on Software Engineering*, Oct. 1988
- *Principles of Software Engineering Management* by Tom Gilb (Addison-Wesley, 1988)
- "A Software Technology Evaluation Program" by David Card in *Information and Software Technology*, July/Aug. 1987
- *Software Inspection* by Tom Gilb and Dorothy Graham (Addison-Wesley, 1993)
- "Experience with Inspection in Ultralarge-Scale Developments" by Glen Russell in *IEEE Software*, Jan. 1991
- *Programming Productivity* by Capers Jones (McGraw-Hill, 1986)

Chapter 8

Parting Words

We have known the fundamentals of the software process for years — one has only to read classic texts such as Fred Brooks' *Mythical Man Month* (originally published in the mid-1970s) to see that this is true. Unfortunately, as an industry, we have generally ignored these fundamentals in favor of flashy new technologies promising to do away with all our complexities, resulting in a consistent failure rate of roughly 85%. Seven out of eight projects fail. That is the cold, hard truth. Additionally, this failure rate and embarrassments such as the Y2K crisis are clear signs that we need to change our ways. It is time for organizations to choose to be successful, to choose to follow techniques and approaches proven to work in practice, to choose to follow a mature software process.

A failure rate of roughly 85% implies a success rate of only 15%.
Think about it.

8.1 Looking Toward Construction

In a nutshell, the goal of the Elaboration phase is to prepare your project team for the Construction phase. To move into the Construction phase, you must pass the Lifecycle Architecture Milestone (Kruchten, 1999). To pass this milestone you must achieve:

Stability. The vision for your project should be stable — the result of working together with your project stakeholders to identify, document, communicate, and then agree to the requirements for your project. The architecture for your project also should be stable — the result of formulating, modeling, and then proving your architectural vision through the use of technical prototyping. Your technical prototype should visibly show that it addresses the major risks to your project — both technical and business/domain risks identified as part of your requirements and project management workflow efforts.

A realistic project plan. You must have a coarse-grained project plan for the entire Construction phase as well as detailed plan(s) for next one or two iterations. Your project plan must be accepted by your project stakeholders, and for this to happen, it must include sufficient detail to show that you understand how to proceed effectively. Your project plan should be supported by credible details, including the base estimating and scheduling assumptions you made during the planning process.

A decision to proceed. A "go/no-go" decision should be made at the end of the Elaboration phase to determine whether it makes sense to continue into the Construction phase. To make this decision, you will likely compare the actual expenditure to date with the expected expenditures to provide insight regarding the accuracy of your estimation. If you found that your estimation was off by $x\%$ for the Elaboration phase, then you should assume that your estimation will once again be off by $x\%$ for the Construction phase — important information to factor into your decision to proceed. The results of your architectural prototyping efforts will also indicate whether or not your approach is technically feasible (if the technology doesn't work, you probably shouldn't proceed). Finally, at some point, you need to assess whether your organization will be able to operate your system once it is in production (if you can't keep the system running once you've built it, there isn't much value building it in the first place).

Software development, maintenance, and support is a complex endeavor — one that requires good people, good tools, good architectures, and good processes to be successful. This four-volume book series presents a collection of best practices for the enhanced lifecycle of the Unified Process published in *Software Development* (www.sdmagazine.com) that were written by luminaries of the information industry. The adoption of the practices that are best suited to your organization is a significant step towards improving your organization's software productivity. Now is the time to learn from our past experiences. Now is the time to choose to succeed.

The Zen of Software Process

User, developer, manager
all peer into the darkness.
Light glimmers in the distance. A solution?
The polar bear does not know the flamingo.

Appendix A

References and Recommended Reading

Printed Resources

Ambler, S.W. (1995). *The Object Primer: The Application Developer's Guide To Object Orientation*. New York: SIGS Books/Cambridge University Press.

Ambler, S.W. (1998a). *Building Object Applications That Work: Your Step-By-Step Handbook for Developing Robust Systems with Object Technology*. New York: SIGS Books/Cambridge University Press.

Ambler, S. W. (1998b). *Process Patterns — Building Large-Scale Systems Using Object Technology*. New York: SIGS Books/Cambridge University Press.

Ambler, S. W. (1999). *More Process Patterns — Delivering Large-Scale Systems Using Object Technology*. New York: SIGS Books/Cambridge University Press.

Ambler, S.W., ed. (2000a). *The Unified Process Construction Phase*. Lawrence, KS: R&DBooks, CMP Media Inc.

Ambler, S.W., ed. (2000b). *The Unified Process Inception Phase*. Lawrence, KS: R&DBooks, CMP Media Inc.

Bassett, P. G. (1997). *Framing Software Reuse: Lessons From the Real World*. Upper Saddle River, NJ: Prentice-Hall, Inc.

Baudoin, C., and Hollowell, G. (1996). *Realizing the Object-Oriented Life Cycle*. Upper Saddle River, New Jersey: Prentice-Hall, Inc.

Beck, K. and Cunningham, W. (1989). *A Laboratory for Teaching Object-Oriented Thinking*. Proceedings of OOPSLA'89, pp. 1–6.

Beck, K. (2000). *Extreme Programming Explained — Embrace Change*. Reading, MA: Addison Wesley Longman, Inc.

Bennett, D. (1997). *Designing Hard Software: The Essential Tasks*. Greenwich, CT: Manning Publications Co.

Binder, R. (1999). *Testing Object-Oriented Systems: Models, Patterns, and Tools*. Reading, MA: Addison Wesley Longman, Inc.

Booch, G. (1996). *Object Solutions — Managing the Object-Oriented Project*. Menlo Park, CA: Addison Wesley Publishing Company, Inc.

Booch, G., Rumbaugh, J., & Jacobson, I. (1999). *The Unified Modeling Language User Guide*. Reading, MA: Addison Wesley Longman, Inc.

Buschmann, F., Meunier, R., Rohnert, H., Sommerlad, P., & Stal, M. (1996). *A Systems of Patterns: Pattern-Oriented Software Architecture*. New York: John Wiley & Sons Ltd.

Champy, J. (1995). *Reengineering Management: The Mandate for New Leadership*. New York: HarperCollins Publishers Inc.

Chidamber S.R. & Kemerer C.F. (1991). *Towards a Suite of Metrics for Object-Oriented Design*. OOPSLA'91 Conference Proceedings, Reading MA: Addison-Wesley Publishing Company, pp. 197-211.

Coad, P. and Mayfield, M. (1997). *Java Design: Building Better Apps and Applets*. Englewood Cliff, NJ: Prentice Hall.

Compton, S.B. & Conner, G.R. (1994). *Configuration Management for Software*. New York: Van Nostrand Reinhold.

Constantine, L. L. (1995). *Constantine on Peopleware*. Englewood Cliffs, NJ: Yourdon Press.

Constantine, L.L. & Lockwood, L.A.D. (1999). *Software For Use: A Practical Guide to the Models and Methods of Usage-Centered Design*. New York: ACM Press.

Coplien, J.O. (1995). *A Generative Development-Process Pattern Language.* Pattern Languages of Program Design, Addison Wesley Longman, Inc., pp. 183-237.

DeLano, D.E. & Rising, L. (1998). *Patterns for System Testing.* Pattern Languages of Program Design 3, eds. Martin, R.C., Riehle, D., and Buschmann, F., Addison Wesley Longman, Inc., pp. 503-525.

DeMarco, T. (1997). *The Deadline: A Novel About Project Management.* New York: Dorset House Publishing.

Emam, K. E.; Drouin J.; and Melo, W. (1998). *SPICE: The Theory and Practice of Software Process Improvement and Capability Determination.* Los Alamitos, California: IEEE Computer Society Press.

Fowler, M. (1997). *Analysis Patterns: Reusable Object Models.* Menlo Park, California: Addison Wesley Longman, Inc.

Fowler, M. and Scott, K. (1997). *UML Distilled: Applying the Standard Object Modeling Language.* Reading, MA: Addison Wesley Longman, Inc.

Gamma, E.; Helm, R.; Johnson, R.; and Vlissides, J. (1995). *Design Patterns: Elements of Reusable Object-Oriented Software.* Reading, Massachusetts: Addison-Wesley Publishing Company.

Gilb, T. & Graham, D. (1993). *Software Inspection.* Addison-Wesley Longman.

Goldberg, A. & Rubin, K.S. (1995). *Succeeding With Objects: Decision Frameworks for Project Management.* Reading, MA: Addison-Wesley Publishing Company Inc.

Grady, R.B. (1992). *Practical Software Metrics For Project Management and Process Improvement.* Englewood Cliffs, NJ: Prentice-Hall, Inc.

Graham, I.; Henderson-Sellers, B.; and Younessi, H. 1997. *The OPEN Process Specification.* New York: ACM Press Books.

Graham, I.; Henderson-Sellers, B.; Simons, A., and Younessi, H. 1997. *The OPEN Toolbox of Techniques.* New York: ACM Press Books.

Hammer, M. & Champy, J. (1993). *Reengineering the Corporation: A Manifesto for Business Revolution.* New York: HarperCollins Publishers Inc.

Humphrey, W.S. (1997). *Managing Technical People: Innovation, Teamwork, And The Software Process.* Reading, MA: Addison-Wesley Longman, Inc.

Jacobson, I., Booch, G., & Rumbaugh, J., (1999). *The Unified Software Development Process.* Reading, MA: Addison Wesley Longman, Inc.

Jacobson, I., Christerson, M., Jonsson, P., Overgaard, G. (1992). *Object-Oriented Software Engineering — A Use Case Driven Approach*. ACM Press.

Jacobson, I., Griss, M., Jonsson, P. (1997). *Software Reuse: Architecture, Process, and Organization for Business Success.* New York: ACM Press.

Jones, C. (1996). *Patterns of Software Systems Failure and Success*. Boston, Massachusetts: International Thomson Computer Press.

Karolak, D.W. (1996). *Software Engineering Risk Management.* Los Alimitos, CA: IEEE Computer Society Press.

Kruchten, P. (1999). *The Rational Unified Process: An Introduction.* Reading, MA: Addison Wesley Longman, Inc.

Lorenz, M. & Kidd, J. (1994). *Object-Oriented Software Metrics.* Englewood Cliffs, NJ: Prentice-Hall.

Maguire, S. (1994). *Debugging the Development Process.* Redmond, WA: Microsoft Press.

Marick, B. (1995). *The Craft of Software Testing : Subsystem Testing Including Object-Based and Object-Oriented Testing.* Englewood Cliff, NJ: Prentice Hall.

Mayhew, D.J. (1992). *Principles and Guidelines in Software User Interface Design.* Englewood Cliffs NJ: Prentice Hall.

McClure, C. (1997). *Software Reuse Techniques: Adding Reuse to the Systems Development Process.* Upper Saddle River, NJ: Prentice-Hall, Inc.

McConnell, S. (1996). *Rapid Development: Taming Wild Software Schedules.* Redmond, WA: Microsoft Press.

Meyer, B. (1995). *Object Success: A Manager's Guide to Object Orientation, Its Impact on the Corporation and Its Use for Engineering the Software Process.* Englewood Cliffs, New Jersey: Prentice Hall, Inc.

Meyer, B. (1997). *Object-Oriented Software Construction, Second Edition.* Upper Saddle River, NJ: Prentice-Hall PTR.

Mowbray, T. (1997). *Architectures: The Seven Deadly Sins of OO Architecture.* New York: SIGS Publishing, Object Magazine April, 1997, 7(1), pp. 22–24.

Page-Jones, M. (1995). *What Every Programmer Should Know About Object-Oriented Design.* New York: Dorset-House Publishing.

Reifer, D. J. (1997). *Practical Software Reuse: Strategies for Introducing Reuse Concepts in Your Organization.* New York: John Wiley and Sons, Inc.

Royce, W. (1998). *Software Project Management: A Unified Framework.* Reading, MA: Addison Wesley Longman, Inc.

Rumbaugh, J., Jacobson, I. & Booch, G., (1999). *The Unified Modeling Language Reference Manual.* Reading, MA: Addison Wesley Longman, Inc.

Siegel, S. 1996. *Object Oriented Software Testing: A Hierarchical Approach.* New York:John Wiley and Sons, Inc.

Software Engineering Institute. (1995). *The Capability Maturity Model: Guidelines for Improving the Software Process.* Reading Massachusetts: Addison-Wesley Publishing Company, Inc.

Szyperski, C. (1998). *Component Software: Beyond Object-Oriented Programming.* New York: ACM Press.

Taylor, D. A. (1995). *Business Engineering With Object Technology.* New York: John Wiley & Sons, Inc.

Warner, J. & Kleppe, A. (1999). *The Object Constraint Language: Precise Modeling With UML.* Reading, MA: Addison Wesley Longman, Inc.

Webster, B.F. (1995). *Pitfalls of Object-Oriented Development.* New York: M&T Books.

Whitaker, K. (1994). *Managing Software Maniacs: Finding, Managing, and Rewarding a Winning Development Team.* New York: John Wiley and Sons, Inc.

Whitmire, S. A. (1997). *Object-Oriented Design Measurement.* New York: John Wiley & Sons, Inc.

Wiegers, K. (1996). *Creating a Software Engineering Culture.* New York: Dorset House Publishing.

Wiegers, K. (1999). *Software Requirements.* Redmond, WA: Microsoft Press.

Wirfs-Brock, R., Wilkerson, B., & Wiener, L. (1990). *Designing Object-Oriented Software.* New Jersey: Prentice Hall, Inc.

Yourdon, E. (1997). *Death March: The Complete Software Developer's Guide to Surviving "Mission Impossible" Projects.* Upper Saddle River, NJ: Prentice-Hall, Inc.

Web-Based Resources

CETUS Links http://www.cetus-links.org

The OPEN Website http://www.open.org.au

The Process Patterns Resource Page http://www.ambysoft.com/
processPatternsPage.html

Rational Unified Process http://www.rational.com/products/rup

Software Engineering Institute Home Page http://www.sei.cmu.edu

Appendix B

The Article Authors

Ambler, Scott W. Scott W. Ambler is the President of Ronin International (www.ronin-intl.com), a firm specializing in software process mentoring and software architecture consulting. He is a contributing editor with *Software Development* and author of the books *The Object Primer* (1995), *Building Object Applications That Work* (1998), *Process Patterns* (1998), *More Process Patterns* (1999), and co-author of *The Elements of Java Style* (2000) all published by Cambridge University Press.

Binder, Robert Robert Binder is president of RBSC Corp., in Chicago, Ill. He is co-author of *The 1989 CASE User Survey*, and the author of *Application Debugging* (Prentice-Hall, 1985) and *Testing Object-Oriented Systems* (Addison Wesley Longman, 1999).

Coad, Peter Peter Coad is the innovator behind the Coad Method, the software development method that bears his name. He founded Object International in 1986 and TogetherSoft in 1999.

Constantine, Larry Larry Constantine is the director of research and development at Constantine and Lockwood Ltd., and Management Forum editor of *Software Development* magazine.

Fowler, Martin Martin Fowler is an independent software consultant specializing in the application of object technology to business information systems. He is the co-author of *UML Distilled* (Addison-Wesley, 1997), and author of *Analysis Patterns* (Addison-Wesley Longman, 1997) and *Refactoring* (Addison Wesley Longman 1999).

Hanscome, Barbara Barbara Hanscome is former editor-in-chief of *Software Development* and is its current associate publisher.

Hargett, Phil Phil Hargett is a consultant with Diamond Technology Partners, a Chicago-based business and information technology consulting firm.

Holmes, Stephen Stephen Holmes is the engineering manager for Lionbridge Technologies, a provider of outsource globalization services, based in Dublin, Ireland.

Linthicum, David S. David S. Linthicum is the chief technical officer of SAGA and author of *Enterprise Application Integration* (Addison Wesley Longman, 1999).

Loomis, Mary Mary Loomis is director of the software technology laboratory of Hewlett-Packard Laboratories in Palo Alto, California.

Maguire, Steve Steve Maguire is a former troubleshooter and project lead for Microsoft Corporation. He has written several books, including *Debugging the Development Process* (Microsoft Press, 1997) and *Writing Solid Code* (Microsoft Press, 1994).

Mayfield, Mark Mark Mayfield is president of JEN Consulting. He helps teams discover abstractions for object models.

McCarthy, Jim Jim McCarthy has participated in the creation and shipping of more than 25 PC software products for companies like Microsoft, AT&T Bell Laboratories, and the Whitewater Group. He is author of the book *The Dynamics of Software Development* (Microsoft Press, 1995).

McClintock, Colleen Colleen McClintock is co-founder of Infinite Intelligence, a company specializing in the application of advanced techniques and technologies to the solutions of complex business problems.

McConnell, Steve Steve McConnell is chief software engineer at Construx Software Builders Inc., a Seattle-area software construction firm. He is the author of the books *Code Complete* (Microsoft Press, 1993) and *Rapid Development: Taming Wild Software Schedules* (Microsoft Press, 1996), the editor of *IEEE Software*, and an active developer.

Rosenberg, Doug Doug Rosenberg is founder and president of ICONIX Software Engineering. He has been teaching Unified Booch/Rumbaugh/Jacobson modeling since 1993. He is the author of the *Mastering UML with Rational Rose* and *Unified Object Modeling Approach* CD-ROM tutorials.

Rothman, Johanna Johanna Rothman provides consulting and training to improve product development practices. She has more than 20 years experience in the software industry.

Szyperski, Clemens Clemens Szyperski is a software architect with Microsoft Research and the author of *Component Software — Beyond Object-Oriented Programming* (Addison Wesley, 1998).

Wiegers, Karl Karl Wiegers is the principal consultant at Process Impact, the author of Jolt Productivity Award-winning *Creating a Software Engineering Culture* (Dorset House, 1996) and *Software Requirements: A Pragmatic Approach* (Microsoft Press, 1999), and a contributing editor to *Software Development*.

Index

Printed and bound by CPI Group (UK) Ltd, Croydon, CR0 4YY

21/10/2024

01777097-0003